THE
SELF-EVOLVING
COSMOS

A Phenomenological Approach
to Nature's Unity-in-Diversity

K&E Series on Knots and Everything – Vol. 18

THE
SELF-EVOLVING
COSMOS

A Phenomenological Approach
to Nature's Unity-in-Diversity

Steven M. Rosen

City University of New York, USA

World Scientific

NEW JERSEY • LONDON • SINGAPORE • BEIJING • SHANGHAI • HONG KONG • TAIPEI • CHENNAI

Published by

World Scientific Publishing Co. Pte. Ltd.

5 Toh Tuck Link, Singapore 596224

USA office: 27 Warren Street, Suite 401-402, Hackensack, NJ 07601

UK office: 57 Shelton Street, Covent Garden, London WC2H 9HE

Library of Congress Cataloging-in-Publication Data
Rosen, Steven M.
    The self-evolving cosmos : a phenomenological approach to nature's unity-in-diversity / by
Steven M Rosen.
        p. cm. -- (Series on knots and everything ; v. 18)
    Includes bibliographical references and index.
    ISBN-13: 978-981-277-173-5 (hardcover : alk. paper)
    ISBN-10: 981-277-173-5 (hardcover : alk. paper)
    ISBN-13: 978-981-283-581-9 (paperback : alk. paper)
    ISBN-10: 981-283-581-4 (paperback : alk. paper)
    1. Physics--Philosophy. I. Title.
    QC6.R56955 2008
    530.01--dc22
                                    2007043868

British Library Cataloguing-in-Publication Data
A catalogue record for this book is available from the British Library.

The image on the cover – courtesy of NASA, ESA, and The Hubble Heritage Team (STScI/AURA)-ESA/
Hubble Collaboration.

First published 2008
Reprinted 2013

Printed in Singapore.

To Marlene, light of my lifeworld

Contents

Illustrations

Figures

Tables

Preface

After hearing a lecture by his younger colleague Wolfgang Pauli, the renowned physicist Niels Bohr is said to have commented: "We are all agreed that your theory is crazy. The question which divides us is whether it is crazy enough to have a chance of being correct. My own feeling is that it is not crazy enough."

Many contemporary physicists acknowledge that the phenomena of their field are so odd, the problems so befuddling to our current ways of thinking, that only a completely "crazy" theoretical approach to them has any possibility of success. What I suggest in the present book is that resolving the problems of modern physics may require something "crazier" still — not just an entirely new theory, but a whole new philosophical base, a new way of intuiting the world. We are going to see that while a spate of "crazy" theories and concepts have been put forward by physicists to account for the fields and forces of nature and their evolution, all theorizing has been tacitly grounded in a set of philosophical presuppositions and postures cast in the classical mold and never opened to question. I intend to show that it is not so much an absence of the right theory that has frustrated physicists' attempts at a comprehensive understanding of the natural world, but the unacknowledged presence of deeply engrained assumptions about that world that are essentially incompatible with the radically non-classical phenomena underlying it.

In subsequent chapters, I will explore in depth the philosophical suppositions of contemporary physics and will develop in detail a proposed alternative. By way of introducing my approach in this preface, I would like to briefly situate it within the recent history of the philosophy of science.

In the first half of the twentieth century, philosophy of science was governed by the analytic tradition of logical positivism. Here "philosophers of science viewed their job as formalizing the methods of science" (Crease 1997, p. 259). However, while positivism still exerts a significant influence on the philosophy of science, in the 1960s voices of dissent began to be heard. Among the earliest was that of Thomas Kuhn (1962), whose historical take on science flew in the face of science's claim to "objective truths" that transcend the vicissitudes of history. After Kuhn's opening initiative, positivist philosophy of science has been questioned in diverse quarters, including the sociology of science, social constructivism, and feminist philosophy. The present work may be located primarily within the challenge to mainstream philosophy of science that has been mounted by *phenomenology*.

As this book unfolds, the ideas and implications of phenomenological philosophy will be extensively explored. For immediate purposes, let us say the following. Positivist philosophy looks for meaning in formally determined relationships among fixed units of knowledge that have been objectively defined. In contrast, phenomenology sees meaning as arising from the hermeneutic (interpretive) interactions of participants in an evolving lifeworld, a dynamic context of lived experience. In an issue of the journal *Man and World* edited by philosopher Robert Crease (1997), the phenomenological-hermeneutic approach to the philosophy of science is laid out systematically. Featured here are the writings of Martin Eger, Eugene Gendlin, Patrick Heelan, Don Ihde, Theodore Kisiel, and Joseph Kockelmans. More recent contributions toward grounding science in phenomenological philosophy include *Models of the Self* (Gallagher and Shear, eds. 2000), *Heidegger's Philosophy of Science* (Glazebrook 2000), *Ideas for a Hermeneutic Phenomenology of the Natural Sciences* (Kockelmans 2002), *How Scientific Practices Matter* (Rouse 2003), *Philosophy of Technology* (Scharff and Dusek, eds. 2003), *Continental Philosophy of Science* (Gutting, ed. 2005), and *Science, Understanding, and Justice* (Eger 2006). Also noteworthy is *Hermeneutic Philosophy of Science, Van Gogh's Eyes, and God* (Babich, ed. 2004), a collection of essays honoring Patrick A. Heelan. Let me focus briefly on the work of Heelan, since it is especially relevant to what I am attempting in this book.

Although alternative approaches to the philosophy of science have met with particularly strong resistance in the philosophy of space, and of quantum mechanics (Uchii 1998), Heelan (1983) has pioneered the effort to open these fields to phenomenological investigation. According to Heelan, "a hermeneutical analysis...would go far to throw light on the basic 'mysteries' of the quantum theory" (1997). Following Heelan's lead, in the chapters to come I offer a unique phenomenological interpretation of quantum theory, quantum gravity, and cosmology — one that draws on the late writings of Martin Heidegger and Maurice Merleau-Ponty, and employs topological imagination in a reflexively intuitive way. At one point, Heelan remarks that "we do not ask of a philosophy that it contribute to the successful practice of science" (1997). In commenting thus, he appears to be implying a disjuncture between the work of philosophy and that of science. Yet is not science philosophy-laden in the sense that its practices are affected by the philosophical assumptions upon which it rests? Has not modern science assumed the Cartesian attitude presupposing the division of subject and object, and has this not strongly influenced the conduct of science, making detached "objectivity" the order of the day? I intend to demonstrate that, in the case of contemporary theoretical physics, scientific progress critically depends on shifting from the stance of Cartesian philosophy to a phenomenological posture that surpasses the subject-object split.

A notable difference between mainstream and phenomenological approaches to physics is that while the former is inclined to deny its philosophy-ladenness, the latter openly acknowledges it. This recognition of philosophical influence lends itself to the integration of philosophical, theoretical, and practical levels of scientific activity. In the phenomenological physics I will undertake, the doing of science and the doing of philosophy merge into a joint endeavor. This vitalizes the philosophy and makes it generative. Rather than merely explaining, analyzing, or critiquing extant physics from a detached vantage point, the philosophy now contributes something new to the physics that specifically addresses physics' own questions concerning space and time, matter and force. The fusion of science and philosophy embodied in the phenomenological physics I shall unfold leads to a *new* physics, with new solutions to long-standing problems that have proven intractable when approached in the conventional manner.

To prevent semantic confusion, let me point out that the term "phenomenological" is already widely used in physics, though with a meaning that differs markedly from the one given here. In his introduction to the phenomenological movement in philosophy, Herbert Spiegelberg (1982) identifies a variety of meanings associated with the word "phenomenological," several of these being "extra-philosophical" (see pp. 7–11). This is the sense in which the adjective "phenomenological" is commonly employed in contemporary physics. "'Phenomenological' laws are understood to be generalizations which simply *describe* regularities in physical events of various types, without regard to their *explanation* or derivation" (Willard n.d.). This descriptive way of doing physics normally involves less technical analysis and mathematical rigor than do formal approaches to the field. The phenomenological approach adopted in the present volume is also less technical and quantitatively exacting than is formal mathematical physics. But unlike the phenomenology widely practiced in physics today, the phenomenological physics of this book is an intuitive enterprise that takes as its point of departure the philosophical insights of thinkers like Merleau-Ponty and Heidegger.

The Self-Evolving Cosmos is intended for philosophically-oriented, interdisciplinary readers drawn to current developments in physics and cosmology. But it may not be enough to say that this work is interdisciplinary. "Transcultural" may be a better term, in the general sense of C. P. Snow's "two cultures." Snow (1959) commented on the regrettably deep division between the sciences and humanities, the latter including literature, art, and traditional philosophy. Doesn't the philosophy of *science* bridge the cultural divide? Mainstream philosophy of science clearly does not, since it has come down squarely on the side of the natural sciences. I venture to say that the approach to philosophy of science offered in this book — submitted in the spirit of Patrick Heelan and other "dissident" philosophers of science (see Chapter 11) — does contribute to spanning the cultural gap in that it exposes the "soft" phenomenological core of the "hardest" of the "hard" sciences, viz. physics. My hope is that readers will share my sense of the significance of the project to the extent of being willing to take up the transcultural challenge. Although the background and training of such readers may bring them to science from the outside, they will be disposed to extending themselves across the cultural chasm. By the same token, transcultural readers initially socialized *within* the language system of science will not be dismayed by the "foreign dialect" they may hear but will be inclined to listen in a new way.

*

The opening chapter of the book situates the development of physics within the context of human development as a whole. Both are seen to involve the quest for unity and individuation. After examining in Chapter 2 the obstacle to unification under the prevailing orientation of contemporary physics, the phenomenological alternative is introduced in Chapter 3. Here we find that the very goal of physics shifts from a quest for static unity to the exploration of a dynamic unity-in-diversity. In Chapters 4 and 5, the *topological* nature of phenomenological physics is considered and a family of primary topodimensional structures is described. Far from being isolated physical objects, these wave-like dimensionalities are seen to constitute whole psychophysical lifeworlds. The next two chapters take up the *evolution* of lifeworld dimensions. In Chapter 6, I employ the idea of symmetry to articulate some essential principles of dimensional transformation and stages of dimensional development. Chapter 7 details the precise pattern by which dimensional vortices evolve in relation to one another. The principles of topodimensional development are applied to physics and cosmogony in the next three chapters. Chapter 8 features a phenomenological rendition of the extra-dimensional (Kaluza-Klein) approach to force-field unification. In Chapter 9, phenomenological intuition is specifically brought to bear on the question of how the universe evolves. The account of cosmic evolution is completed in the following chapter, where the generation of fundamental force fields and matter particles is worked out in detail via a cosmodimensional matrix. The book concludes with a chapter that highlights the *psychophysical* nature of cosmogony. Psychical aspects of all fundamental particles are identified and the reflexive character of phenomenological physics is explored.

The Self-Evolving Cosmos is the culmination of work dating back to the 1970s when I first applied intuitive topology to physics and other natural sciences. These efforts were carried forward in a series of essays later published as a collection under the title *Science, Paradox, and the Moebius Principle* (1994). One essay in particular paved the way for the present work. Titled "A Neo-Intuitive Proposal for Kaluza-Klein Unification," this paper is a preliminary attempt to address physics' problem of unified field theory via a phenomenological use of topology. Though problems in theoretical physics were subsequently examined in my *Dimensions of Apeiron* (2004), the question of unifying the forces of nature was not systematically engaged. In my more recent volume, *Topologies of the Flesh* (2006), topological phenomenology is advanced by working out the details of basic topodimensional processes and their co-evolution, but no attention is given here to issues in physics or cosmology. Building on these earlier initiatives, the present volume offers a full-blown phenomenological rendering of nature's unity-in-diversity in our self-evolving cosmos.

I would like to acknowledge the encouragement and support I have received from a number of individuals in the course of preparing this book. I am gratefully indebted to Chris Alvarez, John Dotson, Lloyd Gilden, Marketa Goetz-Stankiewicz, Brian D. Josephson, Yair Neuman, Ronald Polack, David Roomy, Marlene Schiwy, Ernest Sherman, Louise Sundararajan, Geo N. Turner, and John R. Wikse. Much appreciated is Series Editor Lou Kauffman's receptive response to this project, and the helpful attention of Editor E. H. Chionh in the production phase. For her patient and meticulous assistance in formatting the manuscript for publication, I give my thanks to Shelley MacDonald Beaulieu. Thanks also go to Lisa Maroski, whose diligent work on the index, eagle's eye for typos, and contagious enthusiasm have been a boon. And I want to thank Martin Gardner and Paul Ryan for their kind permission to use their topological drawings in Chapter 4 of this book.

Chapter 1

Introduction

Individuation and the Quest for Unity

The quest for unity in science may be traced to the oldest, most basic of human drives: that toward *individuality*. To be "in-dividual" is to be undivided, to possess a unitary core, a coherent and stable center of identity. From its inception, human development is guided by the tendency toward individuation.

In Sigmund Freud's approach, the emergence of internal unity is reflected in the infant's "primary narcissism" (1914), the diffusely cohesive image of its own body that constitutes the earliest manifestation of the ego. The dialectical nature of the individuation process is brought out in Lacan's (1953) elaboration on Freud. Lacan suggests that the first tentative appearance of an invariant body image occurs in the "mirror stage." Here the child begins to develop a stable sense of its identity, to form an image of its own body as a whole, when the mother's body can be mirrored back to it as its primary object or alter-ego. This possibility initially arises at around six months of age and depends, in turn, on the psychological separation of the infantile body from that of the mother. Developmental theorist Elizabeth Grosz (1994), in drawing on Lacan, thus observes that it is in the mirror stage that "the division between subject and object (even the subject's capacity to take itself as an object) becomes possible for the first time" (p. 32). From the very outset then, one's sense of oneself as an autonomous being is linked to one's experience of the other.

In examining the original condition for the emergence of individuality, Grosz brings in the role played by *space*:

> For the subject to take up a position as a subject, it must be able to be situated in the space occupied by its body. This anchoring of subjec-

tivity in its body is the condition of coherent identity, and, moreover, the condition under which the subject *has a perspective* on the world, and becomes a source for vision, a point from which vision emanates and to which light is focused. (p. 47)

The emergent subject is "the focal point organizing space. The representation of space is...a correlate of one's ability to locate oneself as the point of reference of space: the space represented is a complement of the kind of subject who occupies it" (p. 47). Historically, the space in question, when fully developed, is "the perspectival space that has dominated perception at least since the Renaissance" (p. 48). In sum: "A stabilized body image, ...a consistent and abiding sense of self and bodily boundaries, requires and entails understanding one's position vis-à-vis others, one's place at the apex or organizing point in the perception of space" (p. 48). So, according to Grosz, the self takes form both in relation to others and to space.

Grosz cites Lacan's observation that, in the mirror stage, "'the total form of the body by which the subject anticipates in a mirage the maturation of his powers is given to him only as a *Gestalt*'" (1994, p. 42). Her conclusion is that the "mirror image provides an anticipatory ideal of unity to which the ego will always aspire" (p. 42). Although this "model of bodily integrity" is something "the subject's experience can never confirm," although "the stability of the unified body image...is always precarious" (p. 43), just this imagined unity of the ego or subject is the

precondition...for all symbolic interactions and for an objective or scientific (i.e., measurable, quantifiable) form of space. The virtual duplication of the subject's body, the creation of a symmetry measured from the mirror plane, is necessary for these more sophisticated, abstract, and derivative notions of spatiality. (Grosz 1994, p. 45)

With these words, Grosz intimates a link between science's long quest for unity (symmetry, invariance, continuity) and the even longer search for unity that underlies the whole course of human development.

The connection actually was established years earlier by physicist David Bohm (1965) in his detailed study of the relationship between the notion of invariance in science and the perceptual development of the child. Bohm demonstrated that the kind of activity in which physicists engage when they relate their initially variable empirical observations to invariant mathematical forms can be traced all the way back to the way in which the child's perceptual world crystallizes from the inchoate flux of infancy. Just as the child comes to recognize, for example, that an object momentarily hidden from view is the *same* object when it reappears, the physicist — operating at a vastly greater level of abstraction to be sure — recognizes that certain relationships between physical variables (force and mass, energy and frequency, etc.) remain invariant with changes in space-time coordinates. According to Bohm, the latter is the ultimate extension of the former.

It should be clear that achieving object constancy — whether we are speaking of objects perceived by the child or objects studied mathematically by the physicist — is dialectically coupled with gaining *subject* constancy. On the one hand, because one's imagined body serves as the frame of reference from which all observations are made, an object seen to change with changes in viewpoint or perspective can be taken as the same object only insofar as the observer's body is implicitly sensed as remaining the same. But the subject could not attain coherence and internal stability in the first place if the original chaos of nature did not already lend itself to being ordered. If a core identity is to take shape, if a cohesive individual is to come into being, "Mother Nature" cannot be entirely erratic; she must be able to mirror back to the subject a modicum of regularity and lawfulness. There can be a stable self "in here" only in relation to a world "out there" that offers some measure of consistent patterning. For the objects encountered in the natural environment to be susceptible to being ordered, the environment must place some constraints upon them. The milieu must be bounded or closed in such a way that the objects can properly be contained. Without placing such a limitation upon the objects, their initially unbounded flux could not become channeled into regular patterns. Functioning in this limitative capacity, the environment assumes the character of *space*.

In Grosz's analysis of the individuation process, three principal terms can indeed be identified: subject, object, and space. This tripartite determination has a distinguished history. We can say, in fact, that, in its idealized form, it constitutes the basis of Western philosophy.

*

In the *Timaeus*, Plato states that "we must make a threefold distinction and think of that which becomes, that in which it becomes, and the model which it resembles" (1965, p. 69). The first term refers to any particular object that is discernible through the senses. The "model" for the transitory object is the "eternal object," i.e., the changeless form or archetype. This perfect form is *eidos*, a rational idea or ordering principle in the mind of the Demiurge. Using his archetypal thoughts as his blueprints, the Divine Creator or transcendent subject fashions an orderly world of particular objects and events. As for "that *in* which [an object] becomes," Plato speaks of the "receptacle," describing the latter as "invisible and formless, all-embracing" (1965, p. 70). It is the vessel used by the Demiurge to *contain* the changing forms without itself changing (1965, p. 69). Plato goes on to characterize the receptacle as *space* (1965, pp. 71–72). The Platonic notion of space constituted the seed for a concept that was to come to fruition and play a critical role in post-Renaissance science and mathematics.

Plato's receptacle actually was not entirely changeless. At times it tended toward inhomogeneity, being given to "irrational motion...fleeting potencies and constantly changing tensions" (Graves 1971, p. 71) that made it susceptible to "springing a leak." That is, Platonic space was prone to being ruptured, to losing its continuity. In the course of the next millennium, however, the concept of space evolved. By the time of Descartes, the notion of space had matured into that of a completely homogeneous continuum. Descartes related his continuum to the idea of *extension*.

Consider, as an illustration, the simplified space represented by a line segment. In the Cartesian approach, it is intuitively self-evident that the line, however short, has extension. It must then be continuous: it can possess no holes or gaps in it, since, if the point-elements composing it were not densely packed, we would not have a line at all but only a collection of extensionless points. The quality of being extended implies the infinite density of the constituent point-elements.

Yet, at the same time, intuitive reflection discloses the paradox that the absence of gaps in the continuum not only holds this classical space together but also permits it to be *indefinitely divided*. Without a gap in the line to interrupt the process, there is no obstacle to the endless partitioning of it into smaller and smaller segments. As a consequence, though the points constituting this continuum indeed are densely packed, they are distinctly set apart from one another. However closely positioned any two points may be, a differentiating boundary permitting further division of the line always exists. As philosopher Milič Čapek put it in his critique of the classical notion of space, "no matter how minute a spatial interval may be, it must always be an *interval* separating two points, each of which is *external* to the other" (1961, p. 19).

The infinite divisibility of the extensive continuum also implies that its constituent elements themselves are unextended. Consequently, the point-elements of the line can have no internal properties, no structure of their own. An element can have no boundary that would separate an interior region of it from what would lie on the outside; *all* must be "on the outside," as it were. In other words, the Cartesian line consists, not of internally substantial, concretely bounded entities, but only of abstract boundedness as such (Rosen 1994, p. 92). Sheer externality alone holds sway — what Heidegger called the "'outside-of-one-another' of the multiplicity of points" (1927/1962, p. 481). Moreover, whereas the point-elements of classical space are utterly unextended, when space is taken as a whole, its extension is unlimited, infinite. Although I have used a finite line segment for illustrative purposes, the line, considered as a dimension unto itself, actually would not be bounded in this way. Rather than its extension being terminated after reaching some arbitrary point, in principle, the line would continue indefinitely. This means that the sheer boundedness of the line is evidenced not only locally in respect to the infinitude of boundaries present within its smallest segment; we see it also in the line as a whole inasmuch as its infinite boundedness would be infinitely extended. Of course, this understanding of space is not limited to the line. Classically conceived, a space of any dimension is an infinitely bounded, infinitely extended continuum.

On the classical view, it would be a category mistake to interpret the infinitude of space as a characteristic of what is *object*. Space is not an object but is the "receptacle" of the objects, the changeless context within which objects are manifested. This distinction, initially made by Plato, is reflected in the thinking of Kant, who held that perceptions of particular objects and events are contingent, always given to variation, but that perceptual awareness is organized in terms of an immutable intuition of space. In the words of Fuller and McMurrin, Kant took the position that "no matter what our sense-experience was like, it would necessarily be smeared over *space* and drawn out in *time*[1]" (1957, Part 2, p. 220). Implied here is the categorial separation of *what* we observe — the circumscribed objects — from the *medium* through which we make our observations. We observe objects *by means of* space; we do not observe space. It is within the infinite boundedness of space that particular boundaries are formed, boundaries that enclose what is concrete and substantial. The concreteness of what appears within boundaries is the particularity of the object. In short, an object most essentially is that which is bound*ed*, whereas space is the contextual bounded*ness* that enables the finite object to appear.

The spatial context is what mediates between object and *subject*. The latter (personified by the Demiurge in Plato's *Timaeus*) is the third term of the classical account and corresponds to what is *un*bounded. That an object possesses boundaries speaks to Descartes's characterization of it as *res extensa*, "an extended thing": what has extension will be bounded. In contrast, the subject is *res cogitans*, a "thinking thing." Entirely without extension in space, the subject has no boundaries or parts. As a consequence, it is indivisible. This is etymologically equivalent to stating that the subject is an *individual*. It is before this unbounded subjectivity that bounded objects are cast (the word "object" comes from the Latin, *objicere*, "to cast before"). The crux of classical cognition then, is *object-in-space-before-subject*. The object is *what* is experienced, the subject is the transcendent perspective *from* which the experience is had, and space is the medium *through* which the experience occurs. The relationship among these three terms is that of categorial separation.

Now, classically understood, the tripartite categorial division does not arise empirically; rather, it is taken as existing *a priori*, as always already given. However, even though the trichotomy is regarded as an ontological imperative present from the first, classical thinking can grant that one's *knowledge* of the absolute threefold division does require time to develop. The classical viewpoint therefore can allow that the subject's initial awareness of its individuality is associated with its sense of being "situated in the space occupied by its body" (Grosz 1994, p. 47). It is just that, on the classical view, this identification of subjectivity with embodiment in space merely reflects the subject's immaturity, its ignorance of its true transcendental condition. Yet if the subject, at bottom, is in fact perfectly indivisible thus transcendent of space, and if its objects are completely divisible thus immanent to space, could there be any genuine *interaction* between subject and object? This is of course but another way of stating the old *mind-body problem* that was never quite put to rest in the classical tradition: If mind and body are ontologically divided, how is it possible for them to interact? Assuming that some kind of interaction is undeniable, we appear led to the conclusion that Descartes's emphatic division of mind and body is in fact an idealization that overlooks the reality. While interaction does pose a difficulty for the dualistic species of classical thought, in monistic idealism the problem would appear to be obviated. Here the claim is made that, since only mind is real, since the body is naught but an illusion, in the final analysis mind-body interaction must be illusory. What idealism has never been able satisfactorily to explain, however, is why such an "illusion" would arise in the first place. In lieu of an explanation, we often are advised to accept the ultimate "mystery of it all." Note, moreover, that, if we look beneath the explicit content of what is asserted to its underlying form, we can see that "monistic" idealism is actually another form of *dualism*. Behind the assertion that body "is not real" is the subtler fact of syntax that body "*is*"; negated in overt content, the body is posited in underlying form; the covert effect of such a statement is to *maintain* the body. Mind is posited in the same basic way: "it is real." So, while the content of idealism discounts the body as "mere illusion" and affirms the mind as "real," the *form* of the classical statement, by positing body and mind in stark opposition to one another (one in simple negation, the other in simple affirmation), effectively renders them categorially distinct. In this failed denial of the body's reality, circumvention of the problem of interaction also fails.

From the dialectical standpoint, mind and body — or subject, object, and space — are not taken as pre-existent, fixed, and mutually exclusive categories. Rather, they are seen to develop in intimate relationship to one another. In fact, the dialectical approach that I propose enables us to see how classical thinking itself develops.

On this account, initially there is neither mind nor body, neither subject nor object nor space in any well-differentiated form — only an inchoate flux of embryonic possibilities. In the earliest fragment of Western philosophy, Anaximander referred to this undifferentiated condition as the *apeiron* (see Rosen 2004). Literally meaning "without measure," the old Greek word was variously interpreted as "limitless," "boundless," "indeterminate," or "unintelligible" (Angeles 1981, p. 14). In the proto-scientific discipline of alchemy, the incipient state of affairs was termed *prime matter*: "prima materia, which is the original chaos and the sea" (Jung 1970, p. 9). From the primordial flux, a subject-object mirroring process ensues. That is, quasi-stable objects are differentiated from the chaotic background in relation to an emergent subject before whom the objects are cast. As indicated above, object and subject constancies mirror each other in mutual feedback, thereby enhancing each other. Object and subject thus emerge together from the ever-changing background turbulence and a modicum of unity or invariance arises in them. At the outset, the stability that is realized is highly tentative and the nascent transactions between subject and object are utterly nonlinear. The unity achieved for the object clearly cannot be said to *cause* the subject to be unified, nor is influence transmitted in the other direction, from subject to object. In their still immature, largely undifferentiated relationship, subject and object achieve their unity ensemble, joined inseparably in a recursive mirror play wherein the flow of influence is wholly reversible and cannot be dichotomously parsed. As the differentiation of subject and object advances however, the initial lack of orientation is superseded and an asymmetry sets in. Eventually the action appears to flow in but a single direction: *from* subject *to* object. The subject, functioning as the seemingly exclusive source of agency, divides the object for the purpose of identifying within it a unity (stability, invariance, etc.) that will further enhance the subject's own unity. At this stage of development, the distinctions arising from the dialectical process have hardened into categorial divisions that are now assumed to have existed from the first. Only by virtue of the problem of interaction does the dialectic make its ghostlike presence felt: there is the haunting question of how the subject could exert any influence at all over an object from which it is categorically split. But this does not stop the subject from proceeding with its program of dividing the object so as to bring unity to itself.

In science, the subject proceeds by *analysis*, a word of Greek origin that means a "dissolving, a resolution of whole into parts; *ana*, up, back, and *lysis*, a loosing, from *lyein*, to loose."[2] Or we may say equivalently that "analysis" connotes "a breaking up."[3] The *modus operandi* of science then is to break things up, or — to use the more common manner of speech — it breaks things down. In fact, the process actually entails a *threefold* break-down or division, since the object to be dissected must first be extracted from its context, and since the analyst him- or herself must assume a de-tached stance. In the humanistic program of classical science, the more ef-fectively objects can be parted, the better they can be controlled, manipu-lated, and shaped so as to solidify the unity or integrity of the subject. By dividing the object into its parts, knowledge is gained of how the parts work together in the whole, and this clarification of the functioning of the object as an invariant whole contributes to the wholeness of the subject. Thus, whatever the purpose of the particular research, scientific analysis essen-tially involves a process of division that distills nature's variability into invariant features in the interest of securing the *analyst's* invariance (unity, stability, constancy, etc.). To be sure, the latter does not just refer to the unity of an individual person. In the work of science, personal needs have been sublimated into broader concerns about the welfare of humankind.

The scientific enterprise is well exemplified by the perennial search for the "basic building blocks" of nature. The object is to be analyzed into smaller and smaller components, dissected until we no longer can do so. At this point we will have "hit bedrock," arrived at the fundamental con-stituents of the object, the atoms that compose it. The Greek word "atom" is functionally equivalent to the word "individual": both mean "not divis-ible." What is indivisible is immutable, not susceptible to change. Reach-ing the atomic substrate of nature thus would mean reaching the point where all of nature's variability will have been eliminated. Nature would now be fully controllable. If the object could be manipulated at the atomic level, that of the ultimate "individual," the individuality of the subject would gain its ultimate reinforcement.

In sum, from the child's first tentative steps toward individuation to the sophisticated initiatives of science, the human enterprise has come to be governed implicitly by a fundamental formula: *object-in-space-before-subject*. The prime directive here is that objects be divided so as to secure and enhance the indivisibility of the subject, a task that is to be accom-plished by situating the object within the infinite divisibility that is space. In the pages to follow we will see that — despite all the revolutions that have transformed physics over the past century, the underlying approach has not changed. But we are going to discover that however effectively this way of achieving unity has worked in the past, when it comes to the *ultimate* unification of physics, no longer is that the case. To complete the project of realizing a unified field theory, gravitational and quantum mechanical forces are to be accounted for in an integrated manner. Why has this task proven so difficult? What is it about quantum gravity that makes it so resistant to treatment under the old formula? Unification of the forces in question necessitates operating at an exceedingly minute level of nature. It is at this subatomic level that the analytic subject had hoped to be able to gain complete control over nature's objectivity. Yet we are going to find that it is precisely at this level that the classical problem of *subject-object interaction* — which had been ignorable at larger scales — now no longer can be. Thus we shall see that addressing the ultimate problem of theoretical physics requires that, at the same time, we come to grips with an ultimate *philosophical* problem. In this book, I intend to demonstrate that a radically new, thoroughly dialectical approach is needed to meet the challenge, one that surpasses classical philosophy's threefold division of object, subject, and space in recognition of their intimate entwinement and transpermeation.

Notes

1. I will explicitly address the role of time in due course.

2. *Webster's New Twentieth Century Dictionary*, 2nd ed., s.v. "analysis."

3. *The American College Dictionary*, 1968 ed., s.v. "analysis."

Chapter 2
The Obstacle to Unification in Modern Physics

2.1 Introduction

In the last chapter we found that science's goal of internal unity is advanced by achieving analytic control over the external world. When an object of nature is initially encountered, it will appear to the scientific observer as a more or less undifferentiated whole that is subject to unspecified variations. In analysis, the object under scrutiny is differentiated, broken down into its constituent parts enabling the analyst to understand precisely how these parts operate in the functioning of the whole. Brought into focus in this way, the object is stabilized, its initially undetermined variability now being eliminated or accounted for so as to allow invariant expression of its orderly patterns of action. By thus obtaining knowledge of the object's inner workings, the analyst is better able to predict its behavior and gain control over it. Needless to say, it is not only *particular* objects that are analyzed in science so as to render them invariant but also, classes of objects and their interrelationships. Thus Newton's analysis of the famous falling apple did not give him insight into the motion of that particular object alone, but into the invariant law of gravitational attraction governing the behavior of bodies throughout the universe. (It may seem at first glance that the notion of an "invariant object" entails a category mistake. For, in the classical formula, viz. object-in-space-before-subject, is it not the *subject* that is regarded changeless, whereas the object is susceptible to variation? We must realize however, that, on the classical view, the invariance of the object derives from the ordering activities of the subject, rather than from the object *per se*.)

Until the nineteenth century, the prime exemplar of unity in physics was Newton's universal theory of gravitation. Then, with the work of pioneers such as Oersted, Faraday, and Maxwell, another major unification was achieved. Magnetism and electricity were found to be closely related aspects of the same underlying force of nature. However, in the last two decades of the nineteenth century Maxwell's formulation of electromagnetism was challenged by the experiments of Michelson and Morley. This research raised doubts about the ether field that Maxwell had assumed to be the medium for the propagation of electromagnetic energy. The motionless "ethereal sea" was to serve as the absolute frame of reference for gauging the movements of electromagnetic waves within it. Since the notion of the ether field was the principal embodiment of the idea of classical space, the failure to confirm the former led to questions about the latter that soon precipitated a revolution in physics. While Maxwell's equations could not be shown to be invariant with respect to the classical dimensions of space and time, Einstein demonstrated that their invariance could be established within a new and integrated framework of space-time. In the special theory of relativity, the equations for electromagnetic interaction remain invariant under global (Lorentzian) transformations of four-dimensional space-time coordinates. Special relativity thus accounts for electromagnetic dynamics by allowing the old space and time to vary (to contract and become dilated, respectively) within a new, more abstract context of changeless space-time.

Now, the special theory of relativity was limited to the interaction of systems that are in uniform motion with respect to each other. Ten years after the 1905 appearance of this theory, Einstein unveiled his general theory. By switching from Minkowski flat space to the far more general Riemannian manifold, Einstein could now explain the interaction of systems in non-uniform relative motion. A crucial feature of general relativity was its demonstration of the equivalence of accelerated motion with gravitational effects. This qualified it as a theory of gravitation surpassing Newton's, which was now subsumed as a special case. Mathematically, the relativistic equations for gravitational interaction are invariant under local (Riemannian) transformations of space-time coordinates. What Einstein's general theory could not do was effectively specify a single invariance group containing electromagnetic and gravitational invariances as sub-groups. That is, the theory stopped short of unifying electromagnetic and gravitational forces. Theodor Kaluza (1921) attempted to address this limitation by suggesting that electromagnetism might be expressed in terms of a fifth dimension added to the four dimensions (three space-like, one time-like) that constitute the known universe in Einstein's theory of gravitation. But sixty years were to pass before the extra-dimensional approach to unifying the force field was to receive serious and sustained attention. The delay was occasioned by the need for progress to be completed in an area of modern physics equally as revolutionary as Einsteinian relativity.

At the close of the nineteenth century, just around the time when physicists were digesting the Michelson-Morley findings, another groundbreaking experiment on electromagnetism was being conducted. Max Planck was investigating blackbody radiation, the emission of electromagnetic energy in a completely absorbent medium, a closed cavity that does not reflect light but soaks it up, then discharges the energy internally. Classical theory faced a difficulty here that was on a par with the problem produced by the Michelson-Morley experiment. If the traditional analysis was correct, energy should be transmitted in a smooth and continuous fashion. Yet this assumption leads to the peculiar prediction that, if a non-reflective body is exposed to intense heat, it should radiate an *infinite amount* of energy — a result that clearly is not borne out by empirical observation. Planck responded to the contradiction by boldly amending the underlying classical assumption. He proposed that light, rather than radiating in a smoothly continuous manner, is transmitted in discrete bundles, *quanta*. The introduction of discontinuity into the theory now brought a remarkable correspondence with empirical data. The new quantum theory could predict laboratory findings to a high degree of accuracy by adding just one parameter, h. This is the constant of proportionality that relates the energy (E) of a quantum of radiation to the frequency (v) of the oscillation that produced it: $E = hv$. The numerical value of h is 6.63×10^{-34} joule-seconds. The extremely small value of Planck's constant is consistent with the fact that, in the familiar world of large scale happenings, energy does appear to propagate in a smoothly continuous fashion. It is only when we "look more closely," examining the microscopic properties of light, that we notice its discontinuous, quantized grain.

By the mid-1920s, the quantum mechanical approach to physics had come to the fore and the problem of unifying the forces of nature was presently defined in terms of the quest for a quantum theory that included gravitation. In this context, Oscar Klein (1926) showed that it is possible to write Schroedinger's wave equation in five independent coordinates, thereby demonstrating the basic compatibility of Kaluza's earlier proposal with quantum mechanics. Regarding the question of how the fifth dimension needed to account for electromagnetism could be accommodated in a universe apparently limited to only four dimensions, Klein supposed that the additional dimension could be compactly curved or compressed, hidden at the ultra-microscopic scale of 10^{-35} meter. What is the significance of that tiny magnitude? It is the length set by Planck below which the observable universe goes completely "out of focus," yielding to an all-pervasive uncertainty. In the original Kaluza-Klein account, the unobservable fifth dimension was assumed to be concealed by the spatiotemporal uncertainty associated with the Planck length.

In the years following the extra-dimensional conjectures of Kaluza and Klein, technological advances permitted the construction of particle accelerators enabling the study of two new fundamental forces: the strong force by which atomic nuclei retain their cohesiveness, and the weak force, which mediates radioactive nuclear decay. At the subatomic scale on which these forces operate, gravitation — the weakest of the forces — plays a negligible role. Therefore, in devising abstract gauge theories to unify subatomic forces (including electromagnetism) as members of the same internal symmetry group, physicists could largely disregard the exterior dimensions of space and time more directly relevant in the context of general relativity. During this phase of the quest for unification, development of a Kaluza-Klein theory was not a high priority. But by the late 1970s, weak and electromagnetic forces successfully had been unified by Weinberg and Salam, and an effective theory of the strong interactions formulated (quantum chromodynamics); moreover, the prospects seemed good for a grand unification encompassing all three forces. Now the deferred question of quantum gravity reasserted itself, and this revived interest in a Kaluza-Klein program (Scherk and Schwarz 1975; Cremmer and Scherk 1976). The approach to quantum gravity that is currently attracting the most attention is *string theory* (pioneered in the late 1960s and 1970s by Gabriele Veneziano, John Schwarz, Joël Scherk, Michael Green, and others). Here the four forces of nature are expressed in terms of several basic Planck-scaled force particles which are assumed to be string-like in character and which require ten dimensions for their unification (in the newest version of string theory, known as *M-theory*, eleven dimensions are actually entailed, though the eleventh dimension is not like the other ten; see Chapter 8).

We shall now proceed to explore at a fundamental level the conceptual implications of the recent quest for unity in physics. As we progress, it will become clear that unification cannot adequately be addressed merely as a theoretical problem, that the *philosophical* questions broached in Chapter 1 need to be confronted in any full and effective treatment of the matter.

In the introductory chapter, I adumbrated the way in which the division of subject, object, and space arose from an underlying dialectical relationship in which the three ontological modalities were in fact inseparably intertwined. I ended the chapter by intimating that the riddle of quantum gravity can be solved only through an approach that pays heed to the dialectic. In the following chapters, a dialectical alternative will be spelled out at length. What I intend to demonstrate in the present chapter is that, while modern unification theory indeed may appear to move toward a dialectical way of thinking by seeming to call into question the classical formula of object-in-space-before-subject, in actuality the old formula has been implicitly maintained and this in effect has precluded unification.

2.2 Does Contemporary Mathematical Physics Actually Depart from the Classical Formulation?

2.2.1 Apparent concretization of mathematical physics via symmetry breaking

The notion of invariance that we have discussed is intimately related to the idea of symmetry. A primary strategy of contemporary theoretical physics is to describe the laws of nature in terms of mathematical symmetry. In general, a symmetry is defined when some characteristic of a body or system remains the same despite the fact that a change has been introduced. For example, if a sphere is transformed by rotating it through any angle about its center, its appearance will not change. The sphere therefore can be said to be symmetric under the operation of rotation. This simple notion of symmetry is generalized in group theory, where a variety of mathematical systems can be classified in terms of the groups of transformation under which they remain invariant. Applying the approach to theoretical physics, the laws of physical interaction are described by means of abstract symmetries. Thus, in the framework of special relativity, electromagnetic interaction is said to be symmetric under global transformations of space-time coordinates, and, in general relativity, gravitational interaction is symmetric or invariant under local transformations of coordinates (as noted above). The technique has been especially emphasized in quantum field theory, where invariance in the form of physical interactions among subatomic particles has been studied under transformations of particle properties such as electric charge or strong nuclear charge ("color"). (Note that, whereas the space-time transformations of relativity theory still bear some concrete relationship to the framework within which ordinary human observation occurs, the "interior space" of quantum field theory is purely a mathematical abstraction.)

Now, if the laws of physics can be expressed in terms of abstract symmetry relations, would it not be possible to define more general symmetry groups under which two or more kinds of physical interactions could be subsumed as subgroups? Such an extended application of the idea of symmetry constitutes the basic rationale for unification. The initial contemporary example was mentioned above: the electroweak unification achieved independently by Weinberg and Salam in 1967–68. At first glance, this accomplishment actually may seem to introduce the potential for a fundamental change in philosophical orientation. The Weinberg-Salam breakthrough may appear to cast doubt on the purity of theoretical physics. Let us see why this is so.

Understood most essentially, the "purity" of physics depends on maintaining the threefold categorial division previously discussed — that among object, subject, and space. To reiterate, the object is *what* is experienced, the subject is the detached perspective *from* which the experience is had, and space is the medium *through* which the experience occurs. The classical categories may also be distinguished in terms of the unidirectional flow of influence we have examined. With the subject taken as the primary source of agency, action spreads irreversibly from it to its object, the effect being mediated by the space in which the subject carries out its operations. The subject, regarded classically as in(di)visible, is essentially a *deus ex machina*; it is the unseen, unmoved mover (cf. Aristotle) of (di)visible objects. We have learned that the initially undifferentiated object is divided by the subject for the underlying purpose of bringing unity and order to the chaotic diversity of nature; in this way, the subject brings unity to itself, realizes its potential, gains cognizance of the individuality that, at bottom, it has in fact always possessed. On the classical account, the object alone is transitory, mutable, given to change; both the subject and the spatial "receptacle" through which it operates are taken as inherently changeless.

The symmetries of mathematical physics express most basically the unity the subject has brought to nature's initial variability. To understand the exact role of symmetry in the trichotomous classical formulation, we may take as a model a graphed equation. Consider the simple example of the equation for a parabola, $y = x^2$.

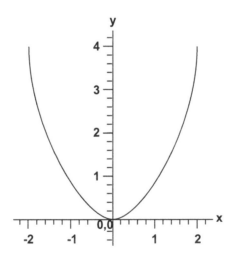

Figure 2.1. Graph of $y = x^2$, the equation for a parabola

In the graph of the equation (Fig. 2.1), x and y are variables, terms whose specific values change from point to point in xy coordinate space. What does not change is the *relationship* between these variables as given by the equation. Since the equation expresses what remains invariant when particular values in the coordinate system are transformed, it can be said to

constitute a symmetry. It is clear that the invariance of the equation depends on the continuity of the functional space in which the relationship
between variables is graphed. Were there a breach in the continuum, at this
singular point the relationship would be abrogated and the equation would
assume a nonfinite value. We can see from this that the symmetry of the
mathematical object (the equation) requires the symmetry of the space in
which the object is contained. Whatever transformations of variables may
occur within the spatial container, for the *relationships* between variables
in the objects under study to be rendered invariant, the container itself
must be invariant; it must stay intact, retain its continuity. Bear in mind
that, on classical thinking, there is a *categorial* distinction between these
two kinds of symmetry. Whereas object symmetry is linked to empirical
observations of variables, the symmetry of space per se is *epistemological*.
The way this was stated in Chapter 1, space is not an object to be known
but is the *means by which* the subject does the knowing.

The subject itself can also be located in the graph of the equation, at
least indirectly. At the center of the xy coordinate system is the 0,0 origin.
The origin of the function space plays a unique role. Representing the locus
at which observation begins, 0,0 serves, in effect, as the surrogate for the
subject's "eyes," the point of view from which its empirical observations are
made (see Rosen 2004, Chapter 1). So the three terms of the classical formula are well discernable in Figure 2.1. The graphed mathematical relationship between variables is associated with the object category; the xy coordinate system is the space in which the object is studied; and the 0,0 origin of
the graph is the subject's point of perspective on that object. In this scheme,
the object symmetry or invariant equation mediates between the variability
of the particular object and the invariance of the subject and its space.

On the classical view, the equations giving the laws of nature are not
themselves empirical objects but more like Plato's "eternal objects" (see
Chapter 1). That is, they are not facts of nature but normative principles
that regulate nature; as such, the laws are not to be thought of as changing
or developing. However, in the contemporary inquiry into the question of
unification, emphasis shifts from concern about relating concretely observable differences to abstract uniformity, to concern about differences
among the abstractions themselves. Although physical processes do not
seem to lend themselves to being expressed in fewer than four distinct

abstract forms, might there nevertheless be a way that these four symmetries could be rendered symmetric with respect to each other, reduced to a single symmetry? The answer proposed by Weinberg and Salam was that these symmetries *were* symmetric with respect to each other in an early phase of cosmogony, but that this all-embracing primordial symmetry subsequently was broken spontaneously. Does this suggestion, now widely accepted by theoretical physicists, imply a reversal of roles between empirical process and normative structure?

In the long-sanctioned approach of attempting to express dynamic occurrences in terms of abstract symmetries, theoretical physicists, in effect, were seeking to relegate concrete change to secondary status, render it an epiphenomenon of changeless structure. But the concept of spontaneous symmetry breaking required for unification weds physics to cosmogony, and, in consummating the marriage, doesn't cosmological time emerge as the measure of formal structure? Would this not amount to a concretization of symmetry relations that previously were regarded as pure abstractions, as normative principles for describing and ordering nature that were themselves outside of nature, exempt from its dynamics? Now it would no longer merely be the *contents* of nature that are viewed as changing in accordance with nature's invariant form; the form itself would be regarded as open to historical processes. Whereas, in classical thinking, said form serves the subject's aim of rendering nature invariant, the dynamization of the laws of nature seemingly implicit in symmetry breaking would call into question the hitherto fixed categorial division between variable empirical objects and the subject's invariant norms; in so doing, it would pave the way for a dialectical approach.

Ten years after the Weinberg-Salam electroweak unification, when the quest for quantum gravity became a pressing issue again, the apparent dynamization of the other hitherto static classical category became evident: in the context of revived Kaluza-Klein theory, space itself was seen to evolve in the course of cosmogony. We know that, in classical physics and philosophy, space and time are taken as the changeless framework within which changes in objects are observed and brought about (by the subject). As Kant saw it, the dimensionality of space and time, its immediately intuited 3+1 character, is no empirical fact but is given *a priori* as an article of pure reason, and thus not susceptible to alteration. These are

the precepts that were challenged in the Kaluza-Klein elaboration on the
Weinberg-Salam idea of spontaneous symmetry breaking.

According to the basic Kaluza-Klein interpretation of cosmogony, the
primordial symmetry condition encompassing all four forms of physical
interaction requires for its expression a compact, multi-dimensional manifold, with all dimensions being "real" ("regarded as true, physical dimensions"; Witten 1981, p. 412), and all coexisting at the same microscopic
scale (close to the Planck length). The event of symmetry breaking is associated with what may be called "dimensional bifurcation": a subset of
dimensions expanded relative to the remaining dimensions, and, in so doing, broke the initial equilibrium. The expanded, 3+1-dimensional universe known to us today — far from being given *a priori* as an article of
pure reason — is the result of this cosmogonic process of dimensional
transformation. Thus, the apparent consequence of the revitalized Kaluza-
Klein approach is that spatiotemporal dimensionality is no longer regarded
strictly as a changeless framework for change. Dimensionality itself is
seemingly thrown into the arena of concrete change, thereby mitigating
the absolute distinction that had been drawn between the spatial container
and the dynamic processes it contains.

But has such a radical step actually been taken in Kaluza-Klein theory?

2.2.2 The implicit attempt to maintain the classical formula and its ultimate failure

Although contemporary Kaluza-Klein theory does render variable what
previously had been taken as invariant, in its mathematical treatment of
this, it actually hopes to stay faithful to the old tripartite formula. For instead of now giving primacy to dynamic process, the new variability described by Kaluza-Klein is to appear in a novel context of superordinate
invariance. Yes, the laws of nature that earlier had seemed to be unchanging are presently viewed as evolving, along with their hitherto changeless
dimensional framework. Yet the evolution is hardly expressed as a dialectical event posing a fundamental challenge to the fixity and separation of
the old ontological categories. Instead the attempt is made to *objectify* the
cosmogonic process by which symmetry is allegedly broken and space-
time transformed; erstwhile invariant laws and their dimensional framework are presently taken as variables in a new invariant equation set that

is to be written for a higher-dimensional, intrinsically invariant space. Therefore, despite the vast difference in the appearance of the Newtonian universe and that of Kaluza-Klein, at bottom both are guided by the underlying principle of object-in-space-before-subject. It is true that the "objects" are now no longer concrete bodies in motion or fixed laws of nature, but laws of nature that evolve, along with their associated spatiotemporal dimensions. Still, this rather more abstract version of nature's objective variability is to be rendered invariant by a detached subject, one who is himself considerably more abstract than the Newtonian subject, and who seeks to operate in a changeless medium of greater dimensionality.

Kaluza-Klein theory is certainly not the first that appears on the surface to dynamize physics while implicitly seeking to preserve the old formula. Kaluza-Klein is an elaboration upon the original theory to adopt this basic strategy: Einsteinian relativity. We have seen that, with Einstein, classical space and time become variables within a new and more abstract four-dimensional framework of changeless space-time. In the Kaluza-Klein account, four-dimensional space-time itself becomes a variable in a still higher-dimensional invariant context.

In the simplified schema for the standard Kaluza-Klein formalism given in Figure 2.2, an initially "compact" two-dimensional space is depicted as expanding to observability along its horizontal axis, its vertical dimension remaining "microcosmically scaled." This differential expansion breaks the purported perfect symmetry of primordial space. By analogy, we may imagine a primordial manifold of ten dimensions (to use the string-theoretic interpretation of cosmogony) being transformed so as to produce our presently observable 3+1-dimensional universe. In this account, the evolving dimensions are of course *objectified*, cast within an analytical continuum or epistemological space that itself does not evolve.

Figure 2.2. Schema for standard formulation of Kaluza-Klein cosmogony

In Kant's day, the categorial distinction between a concrete object and its spatial context was fairly clear-cut. Once again, the object is *what* is observed, whereas space is the *medium* or *means* by which the observation is made. In observing the object, we bring it into focus as a figure that stands out from its spatial background. A fundamental characteristic of the well-focused object is that the boundary separating its interior region from what lies outside of it is clearly demarked. Naturally, without such a boundary, the object could possess no internal structure. Space, for its part, constitutes the context of infinite boundedness within which the object's finite particular boundaries are able to form (see Chapter 1). What happens with the objectification of space itself carried out in modern physics is that the classical distinction between object and space is no longer so simple and straightforward. But by no means is the distinction abolished. Figure 2.2 constitutes an elementary illustration of the continuous transformation of an objectified space (the square expands into a rectangle) that is embedded in an analytical plane that itself goes untransformed. Whereas the objectified space is conceived as possessing topological structure, its analytical context in effect consists only of an infinitude of bounding elements that are themselves structureless.

As of this writing, the Kaluza-Klein unification program has not been carried to successful completion. However, many in the scientific community believe it is only a matter of time before that will happen. From this view I respectfully dissent. If the Einsteinian objectification of space could be worked out successfully, why am I so pessimistic about the prospects for Kaluza-Klein objectification? What Kaluza-Klein theory ultimately requires is the full merging of Einstein's gravitational account with the quantum mechanical rendering of the other forces of nature. And what we are going to see is that gravitational and quantum mechanical forces in fact cannot be unified in a grand symmetric order, as conventional Kaluza-Klein theory would hope, because the primordial environment predating their separation, rather than lending itself to a symmetric depiction, entails a dissymmetry so chaotic that it cannot effectively be repressed. Must we then resign ourselves to ultimate chaos? I suggest the answer is no. I submit that unification of a sort can in fact be achieved. But there is a major proviso. We must be willing to undertake a fundamental reworking of the philosophical foundations of Kaluza-Klein theory. In such a reformulation, the classical trichotomy that implicitly governs the extant approach is to be supplanted by a dialectic.

In order to fully understand what is blocking the unification of general relativity and quantum mechanics, a better understanding of each of these approaches should prove helpful. It turns out that each theory is inherently limited, and that the shortcomings can effectively be ignored or at least tolerated only as long as we attempt no unification. Let us now examine the theories more closely to see where their limitations lie.

Recall that general relativity deals with a more general form of motion than does the special theory, one requiring a manifold of greater mathematical complexity. For this purpose, Einstein turned away from the geometric approach of Minkowski, which was limited to flat space-time, and adopted the more comprehensive curved geometry of Riemann. The flexibility of Riemannian geometry permitted Einstein to gauge the degree of non-uniformity of motion in precise terms by associating it with the degree of curvature in the manifold. Space-time is without curvature for systems in uniform motion and becomes progressively more curved as the acceleration of the inertial frame increases. Applying the principle that establishes the equivalence of inertial and gravitational masses, space-time curvature is related to gravitational effects: the greater the gravitational mass of a body, the more curved is the space-time continuum.

Now, while Einstein found it necessary to adopt this approach, he soon realized that it had its limitations. For, there were solutions to the field equations of general relativity that predicted *infinite* curvature. That is, if a gravitational body were massive enough, the curvature of space-time would become so great that a singularity would be produced in the continuum. What this meant is that analytic continuity would be lost and the theory would fail! However, for that to happen, the mass density of the gravitational body indeed would have to be enormous. When the general theory was first propounded in 1915, the existence of such astrophysical bodies was taken as purely hypothetical. But, as the twentieth century wore on, the possibility of stellar objects whose masses were sufficient to produce "black holes" in space began to be considered more seriously. This led physicist Brandon Carter (1968) to raise explicit doubts about Einstein's theory: Would it be able to survive its prediction of gravitational collapse? By the end of the twentieth century, empirical evidence for black holes had only grown stronger, and, now, as we begin the new millennium, the evidence seems almost irrefutable. One might think that, as a consequence, Einstein's theory might have lost much of its influence. Before considering why that is not the case, let me summarize the theory's course of development and reflect on its meaning.

Einsteinian relativity evolved out of the attempt to circumvent the "black hole" that was created when Michelson and Morley failed to confirm the existence of the luminiferous ethereal continuum. The effect of Einstein's theory was to plug the implicit gap in three-dimensional space by postulating a four-dimensional space-time continuum. To generalize the new account to non-uniform motion, Einstein posited the curvature of space-time. What we are seeing, in effect, is that the four-dimensional approach used to compensate for the absence of continuity in three-dimensional space winds up re-introducing *dis*continuity. Even though general relativity permits one to establish invariances involving non-uniform motion, invariances that presuppose continuity, the *greater* the non-uniformity, the greater is the curvature of space-time, and the closer one then approaches to the point where invariance breaks down and continuity is lost. So it seems that the moment curved Riemannian geometry was applied to generalize Einstein's remedy for discontinuity, a new order of discontinuity was prefigured! More bluntly stated, Einstein's solution *didn't work*; at bottom, it did not effectively address the crisis in contemporary physics precipitated by the Michelson-Morley experiment.

Then why does Einsteinian relativity remain so influential? Should its inherent discontinuity not have undermined it by now? Perhaps an important reason why Einstein's theory has retained its influence is that the other preeminent field of contemporary physics — quantum mechanics (QM) — has created an atmosphere in which discontinuity can better be tolerated and its ultimate consequences better denied.

It took a generation for the truly revolutionary implications of quantum mechanics to become clear. Under the lingering sway of classical thinking, it was natural to assume that the discontinuity of energy found in QM was not really fundamental. For, if the properties of a quantum of energy were to be subject to complete scientific determination, it seemed as if the discontinuity ultimately had to be reducible to continuous expression via an underlying space-time substrate. Yet, by 1930, most physicists had arrived at the conclusion that no such reduction is possible. At this point, the majority of researchers felt obliged to accept the idea that Planck's microscopic quantization implies a basic indivisibility of energy that confounds analytic continuity and, in so doing, calls into question all classical thinking about space and time, including that of Einstein. Therefore, in light of quantum mechanics, "the concepts of spatial and temporal continuity are hardly adequate tools for dealing with the microphysical reality" (Čapek 1961, p. 238).

The microscopic loss of continuity may be better understood by considering more closely Planck's constant, h. This number gives a quantum of *action*. If we rewrite Planck's basic equation, $E = hv$, by replacing frequency (v) with its inverse, namely, time, we then have $E = h/T$ or $h = ET$, and, in physics, the expenditure of energy per unit of time is a measure of action. The *angularity* of quantized action, its internal "spin," is expressed by the application of phase, as given in the formula $h/2\pi = \hbar$. Here h is operated upon by a phase of 2π radians, equivalent to a turn of 360°. In quantum mechanics, $\hbar$ is regarded as an indivisible "atom of process," one not reducible to smaller units that could be applied in its quantitative analysis. Thus, at the sub-microscopic Planck threshold of 10^{-35} meter, the analytical continuity of space gives way to a "graininess" or discreteness that admits of no further quantitative determination. We see here the intimate relationship between the indivisibility of the quantum domain and its basic indeterminacy or uncertainty. According to the uncertainty principle postulated by Werner Heisenberg in 1927, there is a built-in limit to the information we can obtain about the physical properties of quantum systems. This limitation can be stated in terms of Planck's constant: $\Delta p \Delta q \approx \hbar$, where p and q are variables such as position and momentum, or time and energy (variables that are paired or conjugated so as to be essentially indivisible from *each other*). The formula says that the product of the uncertainties (Δs) of such paired terms approximately equals (cannot be less than) the value of Planck's constant. Clearly then, the phasic indivisibility ($h/2\pi$) of Planck-level action is equivalent to its uncertainty ($\Delta p \Delta q$).

There is another way to look at the quantum uncertainty. Nearing the sub-microscopic Planck length, it appears that precise objective measurement is thwarted by the fact that the energy that must be transferred to a system in order to observe it *disturbs* that system significantly. This well-known "problem of measurement" in quantum mechanics expresses quantum indivisibility in terms of the indivisibility of the *observer and the observed*. It seems that, in QM, the observer no longer can maintain the classical posture of detached objectivity; unavoidably, s/he will be an active participant. Evidently this means that quantum mechanical action cannot be regarded merely as objective but must be seen as entailing an intimate blending of object and subject. We will see shortly how the practitioners of quantum mechanics have reacted to the threat to objectivism that has confronted them.

The ultra-microscopic research on light generally appears to hold the same implications for the continuum as did the research of Michelson and Morley: Moving down the scale of magnitude to the Planckian threshold, there is a confrontation with discontinuity. The classical expectation, rooted in the assumption of continuity, is that scale shrinkage eventually brings us to the null volume of the dimensionless point. At this level of nature, all material bodies lose their extension and vanish. As a consequence, space becomes vacuous and cold, devoid of energy or matter. Quantum mechanics presents a dramatically different picture of microphysical reality. Instead of energetic nature being totally subdued, she returns with a vengeance. As the philosopher-physicist David Bohm put it, in classical physics, "the field changes over very short distances are negligibly small," whereas, "in the quantum theory…the shorter the distances one considers, the more violent are the quantum fluctuations associated with the 'zero-point energy' of the vacuum. Indeed, these fluctuations are so large that the assumption that the field operators are continuous functions of position (and time) is not valid in a strict sense" (1980, p. 85).

The vacuum fluctuations that disrupt classical continuity are of course associated with Planck's constant, which embodies the indivisibility of action, irreducible uncertainty, the inseparability of subject and object. This wild variability of nature in the small is clearly reminiscent of the "irrational motion," the "fleeting potencies and constantly changing tensions" (Graves 1971, p. 71) of Plato's tenuously continuous receptacle (see Chapter 1). We have seen that, with the Renaissance, Platonic protospace was superseded by a new order of space, one whose continuity was enhanced. Just as this order was subsequently thrown into doubt by the Michelson-Morley experiment on the velocity of light, so too was it called into question by Planck's blackbody research. Could we say that, whereas

Einstein attempted to plug the hole in the spatial vessel by denying it (via his proposal of a four-dimensional space-time continuum), Planck and his successors fully *accepted* the discontinuity? Did quantum physics give up Einstein's effort to uphold objectivism? Did it embrace the indivisibility of object and subject? These questions must be answered in the negative. QM certainly did not just relinquish continuity and the objectivity it conferred. Instead, implicitly, the attempt was made to retain continuity through an approach that is even more abstract than Einstein's.

Let us consider a central feature of the quantum theoretic formalism: analysis by probability. According to the classical ideal, the extensive continuum is infinitely differentiable, which means that the position of a system within it is always uniquely determinable. When QM was confronted with the *inability* to precisely determine the position of a particle in microspace, it did not merely resign itself to the lack of continuity that creates this fundamental uncertainty. Instead of allowing the conclusion that a microsystem in principle cannot occupy a completely distinct position — which would be tantamount to admitting that microspace is not completely continuous — a multiplicity of continuous spaces was axiomatically invoked to account for the "probable" positions of the particle — "it" is locally "here" with a certain probability, or "there" with another. This collection of spaces is known as *Hilbert space*. *N*-dimensional Hilbert space plays a role similar to that played by Einstein's four-dimensional space-time continuum: it responds to the threat of discontinuity by restoring continuity through an act of abstraction. And the quantum mechanical abstraction of classical space brings with it an abstraction of subjectivity, as also happens in Einsteinian relativity.

Pre-Einsteinian physics could ignore the observer's local perspective, its local space and time, and assume a global perspective in which the 0,0 origin is implicitly taken as the "totally objective" viewpoint of a universal observer (a "Laplacean demon"). What Einstein discovered is that, when it comes to the phenomena of contemporary physics — phenomena whose velocities approach the velocity of light — the local space and time of the concrete observer could not merely be discounted. Of course, Einstein did not simply *accept* the intrinsic variability of concrete observation. In his revision of classical physics, concrete observation itself is explicitly included in the account of nature by making this subjectivity into a *new object*, one whose transformations are invariant in a higher-dimensional context. Thus, there is a substantial difference between the pre-Einsteinian and Einsteinian versions of the classical posture. In the former, we have objective events occurring in three-dimensional space before the observing gaze of an idealized subject. In the latter, where subjectivity itself is taken as object, our "objects" are *observational* events transpiring in four-dimensional space-time. Whereas three-dimensional events are concretely observable, the fourth dimension of Einsteinian relativity is an abstraction. The higher-order Einsteinian observer of these four-dimensional acts of observation is therefore a further step removed from concrete reality than was his Cartesian predecessor. Nevertheless, in both cases, the traditional stance is strictly maintained. In both, we have object-in-space-before-subject.

Like Einsteinian relativity, quantum mechanics implicitly transforms the old subject into an object cast before the analytic gaze of a more abstract, higher-order subject. In effect, the quantum mechanical analyst assumes a superordinate vantage point from which s/he is able to consider alternative acts of classical observation and weight them probabilistically, with each act corresponding to a different subspace of the Hilbert space. That the QM analyst is not so much observing objectively existing particles but the *observations* of particles, is implicit in the attitude of the founding father of quantum mechanics. Max Planck's departure from classical objectivism is especially evident in the following remarks:

> "We have no grounds on which to make any sort of comparison between the actualities of external effects and those of the perceptions provoked by them. All ideas we form of the outer world are ultimately only reflections of our own perceptions. Can we logically set up against our self-consciousness a 'Nature' independent of it? Are not all so-called natural laws really nothing more or less than expedient rules with which we associate the run of our perceptions as exactly and conveniently as possible?" (Planck quoted in Jahn and Dunne 1984, p. 9)

If it seems that Planck was merely advocating subjectivism in expressing these sentiments, that was certainly not the case. At bottom, Planck was not saying that "all is subjective," any more than Einstein was saying that "all is relative." The idea was not that, in surpassing the old objectivism,

the quantum mechanical subject becomes *fused* with the objects s/he observes. On the contrary, detached objective analysis was still the order of
the day. Only now, the "objects" to be analyzed were not concrete substances but the "run of our perceptions." If the quantum mechanical analysis of perception was not to dissolve into a "viciously circular subjectivism," if "scientific objectivity" was to be maintained, the perceptions of
the *analyst* of perception themselves had to be *exempted* from the analysis.
That is to say, two distinct, categorially separate levels of perceptual activity had to exist: that which was to be analyzed, and that through which the
analysis was to take place. The former was constituted by the old observational activity that was now objectified within the framework of the Hilbert space, whereas the latter corresponded to the more abstract, higher-
order, wholly implicit perceptual activity of the quantum mechanical
"observer" standing outside of Hilbert space. It is clear that this quantum
mechanical subject assumed the same detached, "purely objective" stance
as did his Cartesian predecessor. Still operative in its essential relations
was the basic formula of object-in-space-before-subject.

Just how effectively does Hilbert space preserve continuity and the
abstract objectivism with which it is coupled? Each subspace of the multi-
space expression is made continuous within itself to uphold the mutual
exclusiveness of the alternative positions of the observed particle. Such
subspaces must be disjoint with respect to each other, their unity being
imposed externally, by fiat, rather than being of an internal, intuitively
compelling order. So, in the name of maintaining mathematical continuity,
a rather extravagantly *dis*continuous state of affairs is actually permitted in
the standard formalism for quantum mechanics, an indefinitely large ag-
gregate of essentially discrete, disunited spaces. This approach is reminiscent of an earlier stratagem for resisting change.

Bohm (1980) describes the universally sanctioned response given at a
time in history when the order of thinking and perceiving that pre-dated
the Renaissance became suspect. The order of ancient Greece was predicated on the idea of the circle, expressed specifically in the conviction that
celestial bodies trace perfectly circular orbits. "To be sure," noted Bohm,

> when more detailed observations were made on the planets, it was
> found that their orbits are not actually perfect circles, but this fact was

accommodated within the prevailing notions of order by considering the orbits of planets as a super-position of *epicycles*, i.e., circles within circles. Thus one sees an example of the remarkable capacity for adaptation within a given notion of order, adaptation that enables one to go on perceiving and talking in terms of essentially fixed notions of this kind in spite of factual evidence that might at first sight seem to necessitate a thorough-going change in such notions. (1980, p. 112)

I might add that, in the Ptolemaic notion of "epicycles," there was a kind of mockery of the image of the circle, an extravagantly complex, wholly gratuitous replication of this image, one that did not truly achieve what it sought, given the fact that the circle actually had lost its original meaning and effectuality. Yet those who were determined to uphold circularity blinded themselves to this.

Do not the self-continuous subspaces of the Hilbert formalism mock continuity in largely the same way the self-circular epicycles of antiquity mocked circularity? The challenge to the status quo that arose in the late nineteenth and early twentieth centuries was formidable indeed. What was being called into question in the revolutionary research on electromagnetism was not merely an objectified image that the subject can reflect upon, but the reflective posture itself, the whole relation of object-in-space-before-subject. And the response to this profound questioning has been decidedly Ptolemaic. In the method that has been adopted, rather than offering a novel approach that would be genuinely amenable to the discontinuous, non-objective phenomena of modern physics, the attempt is made to pour the "new wine" into an "old bottle." The standard stratagem is to express the upsurgent discontinuity in such a way that continuity is implicitly preserved. Yet it is evident that continuity actually is not successfully maintained in quantum mechanics; despite the artificial semblance of it, on the subtler level of the form that quantum theory takes, continuity is denied (Rosen 1983, 1994, 2004). (Also relevant to this issue are Bohm's [1980] and Josephson's [1987] critiques of mathematical formalism, and the critiques of Stapp [1979] and Gendlin and Lemke [1983].) As a matter of fact, if we follow the full development of quantum mechanics, we see that, in the end, even the semblance of continuity is lost.

Hilbert space does not retain its quasi-continuity for all levels of energy at all scales of magnification. Its range of applicability is limited to the comparatively low-energy regime that lies above the Planck length of 10^{-35} meter. It is true that, in studying the phenomenon of electromagnetic radiation, Planck brought us into an energy domain in which classical continuity was shaken. But while the domain in question is surely microscopic and Planckian uncertainty becomes a significant factor here (whereas, in the large-scale classical world, it does not), this realm of interaction remains considerably above the *ultra*-microscopic, ultra-energetic Planck scale.

However, in the course of the twentieth century, physicists probed the microworld ever more deeply. This permitted them to advance their project of arriving at a unified understanding of nature, as we have seen. Whereas the fundamental forces of nature appear asymmetric with respect to one another at lower energy levels and orders of magnification, physicists, by pushing their quantum mechanical research into the high-energy, sub-microscopic domain of "primordial symmetry," could now account for the atomic decay force (the weak interaction) and the electromagnetic force in a unified manner. Going still further into the microworld, impressive progress was made on a "grand unification" that incorporated the strong nuclear force. And yet, in drawing closer and closer to the Planck length, the element of uncertainty only grew greater.

To complete its quest for unity, physics now faces one final task. It must include in its quantum mechanical analysis the one force of nature hitherto unaccounted for, namely, gravitation. The problem is that gravity, unlike the non-gravitational forces, resists QM treatment until the bitter end. That is, gravitational energy behaves classically, appears to retain its continuity all the way down the scale of magnitude to the Planck length itself. It is precisely here that a QM theory of gravitation would have to operate to fulfill its aim of total unification. Of course, the Planck length is the threshold at which spatiotemporal turbulence goes out of control and uncertainty becomes all consuming. Crossing this threshold, the quasi-continuity of Hilbert space yields to utter discontinuity. Not even a probabilistic analysis of nature is possible here, as is reflected in the unworkable probability values obtained for equations dealing with sub-Planckian reality.

What we are now better able to see is the intimate relationship between quantum mechanics and Einstein's theory of gravitation. It is not merely that the former reaches its Planckian limit and encounters irrepressible discontinuity in a manner that is *analogous* to the black-hole limitation confronting the latter. For what may seem at first like analogous but different limitations, actually can be said to constitute the very *same* limitation. The work of physicist Arthur Eddington (1946) contributed to an understanding of this. In his own effort at unification, he demonstrated the equivalence of relativistic curvature with quantum mechanical uncertainty:

> Curvature and [quantum mechanical] wave functions are alternative ways of representing distributions of energy and momentum....We have introduced the curved space of molar relativity theory as a mode of representation of the extraordinary fluctuation, and have obtained the fundamental relation (3 · 8) between the microscopic constant σ and the cosmological constants R_0, N. (1946, p. 46)

Bearing in mind the essential equivalence of uncertainty with *discontinuity*, Eddington's findings are consistent with the fact mentioned earlier that, the greater the curvature of space-time, the closer we approach to the loss of continuity realized in the singularity of the black hole. Therefore, the production of curvature in general relativity, which culminates in the infinite warping of space-time found in the heart of a black hole, maps onto the production of Planckian uncertainty, the degree of which progressively increases as we descend into the microcosmos. It seems then that the black hole singularity of general relativity is none other than the manifestation of quantized gravitational energy at the Planck length.

It is true, however, that general relativity and quantum mechanics do not converge explicitly until the "bitter end." Before the curvature of space-time actually reaches an infinite value in the black-hole singularity, gravitational uncertainty is only implicit, is not expressed overtly. This limitation can be clarified by considering the question of scale. Are not black holes *large-scale* phenomena, astrophysical events taking place at the opposite end of the scale of magnitude from the microphysical happenings of quantum mechanics? On the contrary, upon entering the singularity of the black hole, the pervasive uncertainty about distance that arises here (owing to the loss of continuity) renders any notion of "large-scale" vs. "small-scale" inoperative. Simply stated, the scale of magnitude collapses. However, so long as the curvature remains finite, its increasing value does not show itself as an out-and-out collapse of scale, but only as a *reduction* of scale accompanying the collapse of *matter*. As a massive stellar body grows older and runs out of hydrogen fuel, the balance of energy is tipped toward gravitation and the attraction of the star's constituent particles for each other gains supremacy, causing the star to implode, to cave in upon itself. This high-density compression of the star leads to a drastic reduction in its volume. The effect is enhanced as we move toward the center of the black hole. Compression becomes so great that we eventually reach the ultra-microscopic Planck threshold. This is the point below which space-time curvature does become infinite and the singularity appears. It is not until we cross this threshold that the microscopically scaled phenomenon becomes one that is utterly unscaleable. Only now is the gravitational uncertainty explicitly manifested, making its presence felt in such a way that it cannot be controlled.

Einstein had hoped to steer clear of microphysical uncertainty. "God does not play dice with the universe," were his famous words. But his hope proved to be vain. The crisis besetting Einstein's theory is that its end result implies a radically discontinuous "microstructure" that blatantly contradicts its initial premise of an entirely continuous one. Einstein's "macroscopic" theory comes to an end at the very same place where quantum mechanics ends: at the Planckian limit at which the quantum of action gains full sway. The singularity at the center of the black hole — far from being a "macroscopic" phenomenon that is merely *analogous* to "microscopic" uncertainty — is that self-same uncertainty. In the final analysis, we cannot distinguish between "large-scale" and "small-scale" forms of uncertainty simply because uncertainty entails the collapse of scale.

Thus, in the singularity of the black hole, relativity theory and quantum mechanics come together. Of course, this "unification of the field" is hardly what science had intended, since the unity is realized in negation, marking as it does the failure of determinative analysis. It is true that, as quantum mechanical research advanced through the twentieth century, a measure of tolerance for uncertainty was built up among physicists. This is a prime reason why Einstein's theory was not simply abandoned with the late twentieth century confirmation that black holes do exist. Had Einstein himself survived to witness the confirmation, it is unlikely he would have been so blasé, for uncertainty was completely anathema to him, even in the smaller doses with which quantum mechanics had inoculated itself. We have seen that Einstein's theory, like quantum mechanics, was essentially an attempt to mitigate the threat of discontinuity brought on by the phenomenon of light. Einstein's way of doing this was to accommodate electromagnetism by extending the classical continuum. His attitude was that continuity must be preserved in a relatively concrete way, without too great a compromise to classical intuition. Yes, the old notion of three-dimensional space plus time now did have to be replaced by the four-dimensional space-time continuum. Nevertheless, the intuition of a unitary geometric continuum was to be kept intact. Quantum mechanics, for its part, did not seek to dispel discontinuity in this concretely intuitive fashion but gave itself the license to proceed *counter*-intuitively; it *allowed* microphysical discontinuity, though formalistically imposing upon it an abstract continuity, as noted above. Therefore, whereas Einstein was completely repelled by discontinuity, quantum mechanics came to take it with greater equanimity.

Yet there is clearly a limit to QM's tolerance of discontinuity. The line is drawn at the Planck length, the threshold below which discontinuity becomes totally unconstrainable. And this is the line at which physics stands today. To complete its "theory of everything," its unified account of the diverse forces of nature, it must now come to terms with gravitation, and this brings it to the Planck length — to the brink of utter chaos, that is. Yet by no means is science prepared to allow its long drive toward unification to be thwarted just when "victory" appears to be in sight. Emboldened by its past success at managing uncertainty, intoxicated by the heady prospect of taming Mother Nature once and for all, contemporary physics seeks ever new techniques for achieving its rational unification, one in which continuity is ultimately retained. A prime example of this is found in string theory. Since this theory has excited widespread interest in the physics community and has resulted in a prodigious amount of research, I will consider it in some detail.

The basic idea is: *What we can't know can't hurt us.* That is, if it is impossible for us to *probe* the cataclysmic energies allegedly prevailing below the Planck length, then we are entitled to assume that said energies either actually do not exist, or that they exist but cannot influence us, and, therefore, can be completely disregarded. The physicist Brian Greene — in his attempt to explain "how string theory calms the violent quantum jitters" (1999, p. 158), "tames the sub-Planck-length quantum undulations of space" (p. 157) — acknowledges the positivistic nature of this strategy: "A positivist would say that something exists only if it can — at least in principle — be probed and measured" (p. 156). Because "the violent sub-Planck-length...fluctuations cannot be measured...according to string theory, [they] do not actually arise" (pp. 156–57).

But how can we say that there is no sub-Planckian world, or that this world does not affect us? Do not the infinities that arise in equations seeking to unify quantum mechanics and general relativity attest to the influence of the ultra-microscopic energies? String theory would answer in the negative. It would say that the troublesome infinities simply come from the pre-string theoretic *assumption* of ultra-microscopic energies. The equations that had been written to successfully unify the three non-gravitational forces had implied that elementary particles are point-like; having no *minimum* length, in principle they could be probed below the Planck length. It is when this assumption is maintained in the attempt to include gravity that we are brought to the sub-Planckian scale where the equations go infinite. To remove the obstacle to formulating a successful theory of quantum gravity, string theory offers an alternative assumption about the basic constituents of the universe. Instead of being unextended, point-like particles, they are *strings*: "tiny, one-dimensional filaments somewhat like infinitely thin rubber bands, vibrating to and fro" (Greene 1999, p. 136). How long is a string? It is roughly the size of the Planck length. By supposing the string to be the *basic* constituent of nature, string theory precludes any movement *below* the Planck length, into the chaotic realm where infinities would arise.

Now, while Greene is a strong proponent of string theory, he admits that the "cosmological positivism" inherent in this approach (— what we can't know can't exist and therefore can't hurt us —) may seem like nothing more than a trick, a conceptual "sleight of hand....Instead of showing that string theory tames the sub-Planck-length quantum undulations of space, we seem to have used the string's nonzero size to skirt the whole issue completely" (p. 157). Having thus raised the question of whether string theory has any real substance, Greene offers a couple of comments intended to suggest that it actually does. However, in his first remark, he only seems to compound the problem.

Here Greene repeats his point that it is not string theory that lacks substance but the infinities of the pre-string theoretic standard model, which are seen to be "an artifact of formulating general relativity and quantum mechanics in a point-particle framework" (p. 157). But is this critique of the standard model followed by a positive demonstration that string theory, for its part, does substantively correspond to reality? On the contrary, Greene merely informs us that string theory furnishes "new rules" that help us play the "game" of eliminating infinities more effectively than did the rules of the earlier model: "String theory tells us that we encountered these problems [of infinities turning up in our equations] only because we did not understand the true rules of the game; the new rules tell us that there is a limit to how finely we can probe the universe" (p. 157). Then is it just a game after all, one whose rules can conveniently be adjusted to obtain the desired results?

The second comment offered by Greene to demonstrate that string theory cannot be dismissed as mere sleight of hand is that it has proven to be

> consistent with the most basic physical principles such as conservation of quantum-mechanical probability (so that physical objects do not suddenly vanish from the universe, without a trace) and the impossibility of faster-than-light-speed transmission of information....The truly impressive feature of string theory is that more than twenty years of exacting research has shown that although certain features are unfamiliar, string theory *does* respect all of the requisite properties inherent in any sensible physical theory. (1999, p. 158)

Nevertheless, while the theory generally conforms to what would be expected of "any sensible physical theory," this is a far cry from providing a specific set of infinity-free equations for quantum gravity. Thus far, it is only *in principle* that string theory eliminates the infinities found in the standard model. Accordingly, Greene is quick to acknowledge that, since the theory has not yet reached a mature stage of development, the jury is still out on it, despite the great promise he believes it holds for unifying the forces of nature. Moreover, even if string theory *could* arrive at a clear-cut theoretical description of quantum gravitational interaction, it would be impossible to test the theory in a concrete way, since the test would have to be conducted on a scale "some hundred million billion times smaller than anything we can directly probe experimentally [!]" (Greene 1999, p. 212).

Now, Greene emphasizes that "the spatially extended nature of a string is the crucial new element allowing for a single harmonious framework incorporating both [gravitational and quantum mechanical] theories" (1999, p. 136). If string theory had proven itself in a substantive and convincing way, it might seem that there would be little point in questioning this characterization of the string. However, given the incompleteness of the theory and the grave difficulties entailed in its external validation (via experimental observation), it appears reasonable to inquire into the *internal* validity of the idea. Does the notion of a fundamental particle with finite extension really make any sense?

Greene poses the question, "What are strings made of?", and answers it as follows:

> Strings are truly fundamental — they are "atoms," *uncuttable constituents*, in the truest sense of the ancient Greeks. As the absolute smallest constituents of anything and everything, they represent the end of the line....From this perspective, *even though strings have spatial extent* [my emphasis], the question of their composition is without any content. (1999, p. 141)

Is there not a contradiction here? In examining the classical concept of spatial continuity in Chapter 1, did we not find that, to be spatially extended is to be *cuttable*, in fact, infinitely divisible? How then could a string be a fundamental particle, an indivisible ingredient of nature, when it is spatially extended? Perhaps, when Greene speaks of the string as he does, he has some unusual definition of "spatial extension" in mind. If that is the case, we are given no hint of it. Consequently, we are left with the default meaning of "extension," the classical intuition of it that entails the idea of being *divisible*. The string, then, is an indivisible particle that is divisible — a contradiction to be sure. Let us look more closely at why string theory is forced into contradiction.

Within the classical framework of object-in-space-before-subject, the idea of an indivisible, point-like atom or particle is indispensable. In order for the subject to analyze effectively an object in space, s/he must be able to determine what the object is "made of," what its basic constituents are. The common tendency is to reify the atom; we think of it as a "tiny object" contained within the larger object that is constituted by it. But atoms are not "contained" in the same manner as ordinary objects are (matches inside a box, for example). For, there is actually a *categorial difference* between the atomic constituent of an object and the object itself. That is because, by definition, the "atoms" of which the object is composed, being indivisible, are not themselves open to analysis. The idea is not to say what *they* are made of; rather, they are the *means* by which the analysis of the larger object is performed. Therefore, the point-like elementary particles, instead of belonging with the object term of the classical formula, in fact belong with the subject term. The fundamental particle to which an object is reduced is the element upon which the subject's analysis of that object is built.

We can say that the particle is the locus or point-origin of the subject's conceptual analysis — much as the subject's *perceptual* activity finds symbolic expression in function space via the 0,0 origin of the xy coordinate system. In our previous examination (p. 18), the 0,0 center of the continuum was characterized as the locus at which observation begins, as surrogate for the subject's "eyes," as the point from which its perspective upon the objective world opens out. Does the idea of the point *particle* enter this perceptual account, as it does in conceptual analysis? Contemporary physics tells us that we observe ordinary objects by bouncing photons off them, which then reach the retinas of our eyes. In this perceptual activity, the photons serve as elementary particles. It is not they that are perceived; rather, these microscopic "atoms" are the means by which the observer perceives the object. The photon thus plays the role of surrogate for the subject. As the elementary probe through which observation takes place, the photon is the point center of the Einstein-Minkowski coordinate system. In sum, whether we are speaking of the perceptual or conceptual operations that the subject performs upon its objects, the point particle is of central importance.

As Greene notes, the standard model of unification maintains the classical assumption of point particles. Of course, the irreducible uncertainty inherent in Planck's constant exposes a basic limitation of this assumption that had not been evident to classical physics. To restate the Planckian restriction a little differently from the way I put it above, it indicates that, the closer we come to the point-like atomic origin of observation and analysis, the fuzzier that origin becomes. Yet, in *non-gravitational* quantum mechanics, the fuzzing out of the 0,0 origin is tolerable. We have seen that contemporary physicists have been content to speak of microscopic measurements in *probabilistic* terms, thereby acknowledging that observation and analysis cannot be fully determinative activities. The prevailing philosophical attitude toward this constraint is given in the positivistic pragmatism of the Copenhagen Interpretation of quantum mechanics. Although there is indeed a built-in limitation to specifying the physical states of a microsystem (say, the position and momentum of an individual particle), the probabilistic analysis can be used with unprecedented exactitude to determine the behavior of macrosystems (whole ensembles of particles). Since the microsystems in question are so miniscule, their uncertainty can be overlooked in the macroscopic operations of the laboratory. Therefore, using Planck's constant, the work of science can proceed quite effectively. But do we not have to acknowledge a complete loss of precision at the ultra-microscopic level of nature? Should this not be of concern to us? Not necessarily, says the Copenhagen Interpretation. By taking microscopic imprecision into account, a high degree of macroscopic precision can be gained. Since the calculations work, there is actually no need to trouble ourselves with the meaning of the underlying uncertainty. What we don't know isn't hurting us in the practical activity of science that we engage in; therefore, we don't have to think about it.

So it is not only string theory that assumes a positivistic stance. The standard model of unification and the Copenhagen Interpretation of quantum mechanics on which it is based also adopt this posture. However, whereas the latter two can afford their positivism, it seems that string theory actually cannot, since *its* positivism must be bought at the price of self-contradiction.

Positivism becomes exorbitantly expensive when gravitation must be included in the unified account of nature. Given the colossal ultra-microscopic energies required for this, even Ptolemaic Hilbert space collapses, since finite probability values no longer can be assigned to terms that describe the physical states of particles. Below the Planck length then, twilight becomes utter darkness. So it seems that, in the case of quantum gravity, what we don't know *does* hurt us. String theory's attempt to circumvent this problem and maintain positivism as a viable approach leads to the apparent contradiction we have discussed: an elementary particle with finite extension — that is, a point particle that is *not* a point particle. The particle *cannot* be a point if the ultra-microscopic infinities are to be avoided. And yet, the particle *must* be point-like, it must be indivisible, if an "objective" analysis is to be performed. For, if the particle is not elementary, if it does not have an unextended atomic character, analytical activity cannot be centered by a 0,0 origin in any definitive fashion.

The contradiction of string theory may be seen in terms of subject and object. Until the last quarter of the twentieth century, physics had done its utmost to suppress the fusion of subject and object intimated by QM's "problem of measurement." The standard model of unification had still been able to assume that the fundamental particle is simply unextended; therefore, the separation of subject and object could continue to be upheld, at least in probabilistic approximation: the point particle is the 0,0 origin of the subject, and the extended entities that are observed and analyzed by the subject are its objects. But then, around 1975, the need to deal with quantum gravity became a pressing issue. Enter here string theory, with its "extended atom." Unlike the point particle of standard unification theory, the string evidently must be *both* indivisible *and* spatially extended — in effect, *both subject and object*. It is this implication that subverts the posture of object-in-space-before-subject that physics had long sought to maintain in its idealized quest for unification.

What the contradiction at the heart of string theory seems to indicate is that, in the final analysis, the sub-Planckian turmoil cannot be denied. Does this preordain a simple regression into chaos? I suggest that it does not. I propose that the apparent self-contradiction of string theory in fact might entail a certain dialectic of transformation.

In Chapter 1 we saw that the work of science is an outgrowth of humanity's long-term striving for individuation and that the latter has required coming to terms with the *apeiron*: the initial inchoate flux of nature in which subject and object are undifferentiated and indistinguishable. But the *apeiron* was never truly eliminated in this quest for unity. Instead, the chaotic element in nature merely had been relegated to the background via the idealizations of science. In the contradiction of string theory, *apeiron* now enters the foreground in an irrepressible way. Yet I submit that the attendant threat of wholesale regression lies not so much in *apeiron* per se, but in the continuing attempt to deny it via idealization. It is in resisting what actually can be resisted no longer that catastrophe threatens. The ongoing resistance is certainly understandable. To the *cogito* (Cartesian subject) who clings to his detached stance, *apeiron* can only signify the complete loss of individuality. But it is surely not the case that the *cogito* had ever *fully achieved* individuality (his idealized projections to the contrary notwithstanding). What we are going to see is that it is precisely when *apeiron* is *accepted* that the process of individuation can be completed in earnest and unification realized in science. By no means does such an acceptance involve passively resigning oneself to dissolution in the chaotic multiplicity of primal nature. We shall discover that one can actively participate in the dazzling variety of nature without *losing* oneness, and that, in fact, oneness (unity, individuality) can genuinely be achieved *only* through said participation. What is called for here is a *dialectic* of the one and many that favors neither the unity of fixed form nor formless multiplicity. A dialectic of *trans*formation is required.

Were this approach to be taken to the blending of subject and object found in string theory, an unacceptable contradiction could become a fruitful paradox. Elsewhere (Rosen 1988, 1994), I attempted just such a dialectical reformulation of the dimensional underpinnings of the theory. Carrying this out effectively involves nothing less than a radical transformation of the foundations of physics. Specifically, the Baconian enterprise of subduing Mother Nature, imposing unity upon Her from afar, is supplanted by an endeavor in which we seek to understand Nature from within, by participating freely in Her flowing diversity (see Fox-Keller 1985 and Chapter 11). In the dialectical rendering of unified field theory, while unity does not simply dissolve, neither does it prevail in the one-sided terms of the *cogito*. Instead there is a *unity-in-diversity*, a *diversity-in-unity*. Indeed, this is the mark of individuation fulfilled, as we are going to see.

Chapter 3
The Phenomenological Challenge to the Classical Formula[1]

Science recognizes its pre-scientific roots. It rightly calls attention to the ignorance and confusion that prevailed before it came onto the scene; it correctly notes the benefits that were reaped with its emergence. But science also tends to believe that it has transcended history. While it readily acknowledges that its particular methods and findings are many and varied, that they are always subject to revision and refinement, science generally takes for granted that its posture of "objectivity" is beyond reproach. The assumption is that this detached stance gives us privileged access to the truth, and that it does so once and for all, exempting it from the possibility that there could be any constructive modification of it in the future. The uncritical acceptance of the "objective" posture has made it difficult for science to address the problem I indicated in the previous chapter, since, as I have shown, contemporary science's problem with regard to unification raises doubts about that posture. When confronted with the limitation of their approach, most working scientists are not likely even to give serious attention to the question of science's basic epistemic position. Instead, they restrict their search for solutions to the "objective" sphere, dismissing the epistemic issue as a matter of "philosophical speculation" that does not concern them. In other words, scientists typically respond to the current philosophical challenge to subject-object dualism by relegating philosophy itself to the "subjective" side. This ostensive rejection of philosophy in fact tacitly affirms the dualistic philosophy from which science emerged. In so doing, it begs the question of subject-object dualism.

I venture to say that scientists who react in this way are out of step with the deeper implications of their own field. According to Kyoto philosopher Tanabe Hajime (1986), "the essential feature of contemporary science...is that philosophy enters into the content of scientific theories, indeed that science cannot stand on its own ground apart from philosophy and therefore must include philosophy within its own theories" (p. 34). For Tanabe, this inability to effectively exclude "subjective" philosophical considerations from "objective" science reflects the fact that, at bottom, the structure of contemporary science is such that "subject and object stand in a dialectical relationship in the sense that each mediates itself by making the other mediate it. This is the structure of active reality whose essence...consists in 'subject-*qua*-object' and 'object-*qua*-subject'" (p. 34). But modern science resists its own inherently dialectical character and, in Ptolemaic fashion, clings to the old philosophy of dualism. It is this philosophical conflict within science that lies behind the problem of unification we are considering. To resolve the conflict, it seems that science must be regrounded in a new philosophy, one that can accommodate the intimate interplay of subject and object.

In the twentieth century, the classical tradition has been perceptively examined by the proponents of a philosophical initiative known as *existential phenomenology*. I will briefly describe the general features of this approach, and then will focus on two phenomenological concepts that have an immediate bearing on the question before us. I am going to demonstrate that the concept of *depth* put forward by Maurice Merleau-Ponty (1964), and of *time-space* advanced by Martin Heidegger (1962/1972), respond to the challenge of unification in science by offering a new understanding of space, time, and dimensionality.

   The phenomenological movement is rooted in the nineteenth century existentialist writings of thinkers like Kierkegaard, Nietzsche, and Dostoevsky. It takes its contemporary form through the work of its principal figures: Edmund Husserl, Martin Heidegger, and Maurice Merleau-Ponty. In terms of the present volume, phenomenology can be seen most essentially as a critique of the classical trichotomy of object-in-space-before-subject. To the phenomenologist, the activities of the detached Cartesian subject are objectifications of the world that conceal the concrete reality of the *lifeworld* (Husserl 1936/1970). Obscured by the lofty abstractions of European science, this earthy realm of lived experience is inhabited by subjects that are not anonymous, that do not fly above the world, exerting their influence from afar. In the lifeworld, the subject is a fully situated, fully-fledged participant engaging in transactions so intimately entangling that it can no longer rightly be taken as separated either from its objects, or from the worldly context itself. As Heidegger put it, the down-to-earth, living subject is a *being-in-the-world* (1927/1962), a being involved in

> a much richer relation than merely the spatial one of being located in the world....This wider kind of personal or existential "inhood" implies the whole relation of "dwelling" in a place. We are not simply located there, but are bound to it by all the ties of work, interest, affection, and so on. (Macquarrie 1968, pp. 14–15)

It is clear that all three terms of the classical formulation are affected by the phenomenological move. We saw in Chapter 1 that, on the classical account, space is the sheer *boundedness* serving as the medium for the *unbounded* subject's operations upon *bounded* objects. The essential principle here is that of external relationship. That is, objects that are simply external to each other appear within a spatial context of *sheer* externality (the "'outside-of-one-another' of the multiplicity of points"; Heidegger 1927/1962, p. 481) and are operated upon by agents acting from beyond space. In the contrasting phenomenological approach, all relations are *internal*. Notwithstanding the Platonic/Cartesian idealization of the world, in the underlying *lifeworld* there is no object with boundaries so sharply defined that it is closed off completely from other objects. The lifeworld is characterized instead by the *transpermeation* of objects, by their mutual interpenetration, by the "reciprocal insertion and intertwining of one in the other," as Merleau-Ponty put it (1968, p. 138). With objects thus related by way of mutual containment, no *separate* container is required to mediate their relations, as would have to be the case with externally related objects. Phenomenological understanding supersedes the classical relationship of container and contained (of sheer boundedness and the merely bounded). Objects no longer are to be thought of as contained in space like things in a box, for, in containing each other, they contain themselves. At the same time, it must also be understood that, in the lifeworld, there can be no categorial division of object and subject. The lifeworld subject, far from being the disengaged, high-flying *deus ex machina* of Descartes, finds itself down among the objects, is "one of the visibles" (Merleau-Ponty 1968, p. 135), is itself always an object to some *other* subject, so that the simple distinction between subject and object is confounded and "we no longer know which sees and which is seen" (Merleau-Ponty 1968, p. 139). This placement of the subject among the objects is of course no materialistic reduction of it to a *mere* object (an inert lump of matter). Rather, the phenomenological grounding of the subject is indicative of the ambiguous interplay of subject and object in the lifeworld. Generally speaking then, what the move from classical philosophy to phenomenology essentially entails is an internalization of the relations among subject, object, and space.

Now, while natural science has not focused its attention on the nature of subjectivity or objectivity, it has been explicitly concerned with the nature of space and time. The dimensions of space and time thus play a pivotal role as a disciplinary bridge: they are foundation stones of science and are at once of critical importance to philosophy in general. Then let us carry forward our examination of the phenomenological alternative by focusing on the ideas of space and time.

In the previous chapter we saw how the classical conception of space — the concept of the continuum — has been seriously challenged by the *dis*continuous phenomena of contemporary science. In my estimation, the phenomenological response to this challenge has been best articulated by Merleau-Ponty (1964). In his concept of *depth*, he provides us with an account of dimensionality that permits us to understand the limitations of Cartesian space and to surpass them. (Although Merleau-Ponty did not comment explicitly on modern physics in his essay on depth, he explored the topic in a lecture course on nature given a few years earlier; see Merleau-Ponty 1956–57/2003.)

By way of introducing Merleau-Ponty's depth dimension, let us consider once more the traditional dichotomy between the objects contained in space and their spatial container, or, as Plato put it, between "that which becomes [and] that in which it becomes" (1965, p. 69). A visible form "becomes," whereas that "*in* which it becomes" is "invisible and formless" (1965, p. 70). Whatever changes may transpire in the objects that "become," however they may be transformed, the containing space itself does not change. Indeed, for there to be change, there must be difference, contrast, dialectical opposition of some kind. But the point-elements that make up the classical continuum, rather than entailing opposition, involve mere *juxta*position. Unextended and thus devoid of inner structure, the elements of space possess no gradations of depth; no shading, texture, or nuance; no contrasts or distinctions of any sort. Instead of expressing the dialectical interplay of shadow and light, space itself is *all light*, as it were. A condition of "total exposure" prevails for the point-elements of the continuum, since these elements, having no interior recesses, must be said to exist solely "on the outside." All that can be said of the relations among such eviscerated beings is what Heidegger said: the points of classical space are "'outside-of-one-another'" (1927/1962, p. 481). So, rather than actively engaging each other as do the beings that are contained in space, the densely packed elements of the classical container sit inertly side by side, like identical beads on a string.

As a matter of fact, even though the beings that dwell in such a space can be described as "actively engaged," the quality of their interaction is affected by the context in which they are embedded: since the continuum is constituted by sheer externality, the relations among its inhabitants must also be external. Classical dynamics are essentially mechanistic; instead of involving a full-fledged dialectic of opposition and identity wherein beings influence each other from core to core, influence is exerted in a more superficial fashion. According to Bohm (1980), the mechanistic order of influence is one in which entities "interact through forces that do not bring about any changes in their essential natures...[they interact] only through some kind of external contact" (p. 173). We may say then that classical space contains dialectical process in such a way that it externalizes it, divesting it of its depth and vitality.

It is true that, with the advent of Einsteinian field theory, the mechanistic approach of Newtonian physics was superseded. Einstein did contribute to dynamizing space and internalizing the interactions transpiring therein. But his objectivist stance prevented his full liberation from classical science. With Einstein, the classical formula reasserted itself at a higher level of abstraction, as we have seen, and this resulted in the eclipse of concrete dialectical process. Not so in the phenomenological approach of Merleau-Ponty. What Merleau-Ponty demonstrates is that the objective space appearing to contain dialectical process actually *originates* from it.

In his essay "Eye and Mind," Merleau-Ponty emphasizes the "absolute positivity" of traditional Cartesian space (1964, p. 173). For Descartes, space simply is *there*; possessing no folds or nuances, it is the utterly explicit openness, the sheer positive extension that constitutes the field of strictly external relations wherein unambiguous measurements can be made. Merleau-Ponty speaks of,

> this space without hiding places which in each of its points is only what it is....Space is in-itself; rather, it is the in-itself *par excellence*. Its definition is *to be* in itself. Every point of space is and is thought to be right where it is — one here, another there; space is the evidence of the "where." Orientation, polarity, envelopment are, in space, derived phenomena inextricably bound to my presence [thus "merely subjective"]. *Space* remains absolutely in itself, everywhere equal to itself, homogeneous; its dimensions, for example, are interchangeable. (1964, p. 173)

Merleau-Ponty concludes that, for Descartes, space is a purely "positive being, outside all points of view, beyond all latency and all depth, having no true thickness" (1964, p. 174).

Challenging the Cartesian view, Merleau-Ponty insists that the dialectical features of perceptual experience ("orientation, polarity, [and] envelopment") are not merely secondary to a space that itself is devoid of such features. He begins his own account of spatiality by exploring the paradoxical interplay of the visible and invisible, of identity and difference, that is characteristic of true depth:

> The enigma consists in the fact that I see things, each one in its place, precisely because they eclipse one another, and that they are rivals before my sight precisely because each one is in its own place. Their exteriority is known in their envelopment and their mutual dependence in their autonomy. Once depth is understood in this way, we can no longer call it a third dimension. In the first place, if it were a dimension, it would be the *first* one; there are forms and definite planes only if it is stipulated how far from me their different parts are. But a *first* dimension that contains all the others is no longer a dimension, at least

in the ordinary sense of a *certain relationship* according to which we make measurements. Depth thus understood is, rather, the experience of the reversibility of dimensions, of a global "locality" — everything in the same place at the same time, a locality from which height, width, and depth [the classical dimensions] are abstracted. (1964, p. 180)

Speaking in the same vein, Merleau-Ponty characterizes depth as "a single dimensionality, a polymorphous Being," from which the Cartesian dimensions of linear extension derive, and "which justifies all [Cartesian dimensions] without being fully expressed by any" (1964, p. 174). The dimension of depth is "both natal space and matrix of every other existing space" (1964, p. 176).

Merleau-Ponty goes on to observe that primal dimensionality must be understood as *self-containing*. This is illustrated through a discussion of contemporary art, and, in particular, the work of Paul Cézanne: "Cézanne knows already what cubism will repeat: that the external form, the envelope, is secondary and derived, that it is not that which causes a thing to take form, that this shell of space must be shattered, this fruit bowl broken" (1964, p. 180). In breaking the "shell," one disrupts the classical representation of objects-in-space. Merleau-Ponty asks:

What is there to paint, then? Cubes, spheres, and cones...? Pure forms which have the solidity of what could be defined by an internal law of construction...? Cézanne made an experiment of this kind in his middle period. He opted for the solid, for space — and came to find that inside this space, a box or container too large for them, the things began to move, color against color; they began to modulate in instability. Thus we must seek space and its content *as* together. (1964, p. 180)

The work of Cézanne is Merleau-Ponty's primary example of the exploration of depth as originary dimension. The foregoing passage describes Cézanne's discovery that primal dimensionality is not space taken in *abstraction* from its content but is the *unbroken flow* from container to content. It is in this sense of the internal mediation of container and content that Cézanne's depth dimension is *self-containing*.

Merleau-Ponty also makes it clear that the primal dimension engages embodied subjectivity: the dimension of depth "goes toward things from, as starting point, this body to which I myself am fastened" (1964, p. 173). In commenting that, "there are forms and definite planes only if it is stipulated how far from *me* their different parts are" (p. 180; italics mine), Merleau-Ponty is conveying the same idea. A little later, he goes further:

> The painter's vision is not a view upon the *outside*, a merely "physical-optical" relation with the world. The world no longer stands before him through representation; rather, it is the painter to whom the things of the world give birth by a sort of concentration or coming-to-itself of the visible. Ultimately the painting relates to nothing at all among experienced things unless it is first of all "autofigurative."....The spectacle is first of all a spectacle of itself before it is a spectacle of something outside of it. (1964, p. 181)

In this passage, the painting of which Merleau-Ponty speaks, in drawing upon the originary dimension of depth, draws in upon itself. Painting of this kind is not merely a signification of objects but a concrete *self*-signification that surpasses the division of object and subject.

In sum, the phenomenological dimension of depth as described by Merleau-Ponty is (1) the "first" dimension, inasmuch as it is the source of the Cartesian dimensions, which are idealizations of it; it is (2) a self-containing dimension, not merely a container for contents that are taken as separate from it; and it is (3) a dimension that blends subject and object concretely, rather than serving as a static staging platform for the objectifications of a detached subject. In realizing depth, we surpass the concept of space as but an inert container and come to understand it as an aspect of an indivisible cycle of action in which the "contained" and "uncontained" — object and subject — are integrally incorporated.

Have we not previously encountered an action cycle of this kind? In the last chapter, we considered the fundamental "atom of process" that lies at the core of quantum mechanics: $\hbar$, the *quantum of action*. The discontinuity associated with quantized microphysical action bespeaks the fact that this indivisible circulation undermines the classical continuum, and, along with it, the idealized objects purported to be enclosed in said continuum and the idealized subject alleged to stand outside it. We know that it is only through Ptolemaic artifice that microphysical action can be accommodated while maintaining the old trichotomy, and that this stratagem is only effective above the Planck length, where the full impact of quantized action can be avoided (see Chapter 2). In addressing the problem of quantum gravity, however, no longer can we remain safely above the Planck length. And it is at or below the Planck scale that quantized action is simply unmanageable as a circumscribed object contained within an analytical continuum from which the analyst is detached. The action in question entails the *indivisible transpermeation* of object, space, and subject — something utterly unthinkable when adhering to the classical formula. Yet just such a dialectic defines the depth dimension as described by Merleau-Ponty. Broadly speaking, this suggests that, when the problem of quantum gravity can no longer be deferred in the quest for unification, science can no longer conduct its business as usual. Instead, a whole new basis for scientific activity is required, a new way of thinking about object, space, and subject, one cast along the lines of Merleau-Pontean depth.

Now, the phenomena of contemporary physics do not merely raise doubts about the classical conception of space, but about the traditional notion of time as well. On the Aristotelian view, time is constituted by a sequence of point-like, extensionless nows. This understanding of time is not so different from the corresponding view of space. As with the spatial continuum, the temporal continuum is composed of densely packed point elements sitting inertly like beads on a string. These temporal elements are distinguished from their spatial counterparts only by the fact that they are given in linear succession rather than simultaneously. Heidegger observed accordingly that the "dimensionality of time, thought as the succession of the sequence of nows, is borrowed from the representation of three-dimensional space" (1962/1972, p. 14). What is implied in the previous chapter is that the continuity of temporal succession is no less confounded by the fundamental discontinuity of contemporary physics than is the continuity of spatial simultaneity. Approaching the Planck length of quantum mechanics, time as well as space loses its classical cohesiveness.

But Heidegger offers a very different way of viewing time. In his 1962 lecture, "Time and Being," he introduces the notion of "time-space" or "true time." According to Heidegger, far from being point-like, "true time proves to be three-dimensional" (1962/1972, p. 15).

Specifically, Heidegger tells us that, behind the appearance of an extensionless, point-like now is: (1) the *past* — which, though denied to us *as* present, nevertheless makes its presence felt in its own way; (2) the *future* — which is withheld from us in *its* own peculiar manner of making itself present; and (3) the present itself (the "presently present"; see Heidegger 1946/1984, p. 35). Heidegger suggests that these three dimensions of time gather together and are held apart in every moment of lived experience. That is, they extend toward each other, yet, since they never merely collapse into one another to form a monotonic unity, the "space" of their intimate interplay remains open. The three dimensions of time determine each other; there is a "mutual giving to one another of future, past and present" (1962/1972, p. 14):

> [the future,] being not yet present, at the same time gives and brings about what is no longer present, the past, and conversely what has been offers future to itself. The reciprocal relation of both...gives and brings about the present....[Thus] future, past and present...belong together in the way they offer themselves to one another. (pp. 13–14)

Heidegger next observes that,

> With this presencing, there opens up what we call time-space. But with the word "time" we no longer mean the succession of a sequence of nows. Accordingly, time-space no longer means merely the distance between two now-points of calculated time....Time-space now is the name for the openness which opens up in the mutual self-extending of futural approach, past and present. This openness exclusively and primarily provides the space in which space as we usually know it can unfold. The self-extending, the opening up, of future, past and present is itself *pre*spatial; only thus can it make room, that is, provide space. (p. 14; emphasis added)

What Heidegger is saying is that — while the "dimensionality of time, thought as the succession of the sequence of nows, is borrowed from the representation of three-dimensional space" (p. 14) — when temporal dimensionality is thought in its *truth*, as the threefold mutual giving, then it is three-dimensional space that derives from *it*.

Expanding on this, Heidegger asserts:

> What is germane to the time-space of true time consists in the mutual reaching out and opening up of future, past and present. Accordingly, what we call dimension and dimensionality...belongs to true time and to it alone. Dimensionality consists in a reaching out that opens up, in which futural approaching brings about what has been, what has been brings about futural approaching, and the reciprocal relation of both brings about the opening up of openness. Thought in terms of this threefold giving, true time proves to be three-dimensional. Dimension, we repeat, is here thought not only as the area of possible measurement, but rather as reaching throughout, as giving and opening up. Only the latter enables us to represent and delimit an area of measurement. (1962/1972, pp. 14–15)

When Heidegger underscores the mutuality or reciprocity of time's three dimensions, when he asserts that they "belong together," it is clear that the past, present, and future are to be grasped dialectically, not trichotomously. This is similar to the way Merleau-Ponty would have us understand the distinct yet dialectically entwined perspectives of the depth dimension. Since Merleau-Ponty's concept of depth as the "first dimension" does seem to parallel Heidegger's notion of "time-space" or "true time" as primal dimensionality, we may ask whether a direct influence is in evidence here. The answer is yes. Heidegger actually had already adumbrated his notion of time's "three ecstases" (past, present, and future) in *Being and Time* (1927/1962), a work that long predates his 1962 lecture. Although Merleau-Ponty makes no reference to Heideggerian temporality in "Eye and Mind" (originally written in 1960), earlier, in *Phenomenology of Perception*, he quotes Heidegger extensively on this (1945/1962, Chapter III-2). Here Merleau-Ponty arrives at the conclusion that human beings are essentially temporal creatures in Heidegger's sense of the term, and, moreover, that the mutual embeddedness and interpenetration of past, present, and future are inextricably related to the phenomenon of perspective: "Change presupposes a certain position which I take up and from which I see things in procession before me: there are no events without someone to whom they happen and whose finite perspective is the basis of their individuality. Time presupposes a view of time" (1945/1962, p. 411). Of course, the dimension of depth of which Merleau-Ponty will later speak (in "Eye and Mind") is also the dimension of perspective: that wherein "I see things, each one in its place, precisely because they eclipse one another...." (1964, p. 180). (Merleau-Pontean perspective must be distinguished from its Cartesian counterpart, which is an idealization of perspectival experience that glosses over its inherent finitude.) What is the difference between Merleau-Ponty's later treatment of perspective and the earlier one that focuses on temporality as such? To answer this question, a digression is called for. Let us return to our consideration of Einsteinian space-time.

In Einstein's special theory of relativity, space and time are united by replacing the hitherto separate measures of distance and duration with a new fundamental unit, the *space-time interval*. A constraint placed on this unification is the need to distinguish unambiguously space-time intervals that are "space-like" from those that are "time-like." To put it roughly, although every space-time interval indeed combines space and time, any given interval must be "weighted" more by one than by the other. Essentially, two events in Einsteinian space-time are separated in a space-like way if the distance between them is great enough relative to their temporal disjunction that an observer could be present for the onset of both events only by traveling from one to the other at the velocity of light or greater. In Einstein's theory, no material body can travel that quickly. Events are separated by a time-like interval if the time elapsing between them is long enough relative to their spatial separation that an observer may be present for the onset of both events, but could not experience them as occurring simultaneously, since to do so again would require traveling at the prohibited velocity.

Figure 3.1 shows the determination of spatiotemporal relations in Einstein's theory. This representation of space-time is known in physics as an Einstein-Minkowski diagram (since Einstein drew from the work of the mathematician Herman Minkowski in developing it). Incorporating the velocity of light into the theory results in the formation of a pair of light

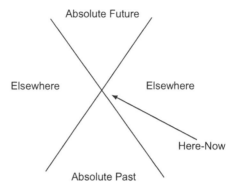

Figure 3.1. Light cones

cones through which time-like and space-like space-time intervals are unequivocally established. The structure of the cones thus sets up a clear-cut dichotomy between events that are successive, and those that are contemporaneous. With respect to the temporal aspect, the cone of the absolute future widens upward from the point of the immediate present ("now"), while the absolute past is extended below the now-point. All events in the future cone are definitely *after* the now, whereas those in the cone of the past are definitely *before* it. This is because the time intervals in question are long enough relative to the spatial separation of events that an observer present for the onset of a given event could indeed be present for the onset of a subsequent or antecedent event, but could not experience either of the latter as occurring *contemporaneously* with the given event, since to do so would require traveling at the prohibited superluminal velocity. In these time-like relations, it is clearly impossible for the future and the past to make their presence felt in the now. Regarding the spatial aspect, the light cones intersect at the point of the immediate *here*. Events outside the cones are definitely elsewhere, as opposed to being here, since the space intervals in question are large enough relative to the temporal separation of events that an observer could be present for the onset of both events only by traveling at the forbidden velocity. In these space-like relations, elsewhere cannot be present here. As can be gathered from the previous chapter, the here-now originating point of Einstein-Minkowski space-time plays a pivotal role in the classical intuition of object-in-space-before-subject. For, as the locus at which observation begins, the origin functions as the surrogate for the subject's "eyes," its point of perspective upon the world, its "window" upon the objects-in-space. The observer's perspective "centers the light cone of a Minkowski diagram, as the point event at its center: (x,t)=(0,0)" (Schumacher 1989, p. 111).

We may apply the distinction between space-like and time-like space-time intervals to Merleau-Ponty's two approaches to perspective. Let us say that, in "Eye and Mind," Merleau-Ponty's primary concern is with the "space-like" aspect of perspectival experience. It is here that he speaks of things "eclipsing" one another so that we cannot simply see both of them at once. From the Einsteinian standpoint, we would view the things in question as perceptual *events*, occurrences whose spatial separation is so great relative to their temporal disjunction (since the events would be taking place approximately at the *same* time) that I would have to travel at the velocity of light or greater to be present for the onset of both. In *Phenomenology of Perception*, Merleau-Ponty is chiefly interested in the "time-like" aspect of perspectival experience and therefore is focused on the question of the relationship of past, present, and future. This is the case of space-time events whose temporal component "outweighs" the spatial.

However, while Einstein's distinction may help us to conceptualize the difference between Merleau-Ponty's two analyses of perspective, it is important to realize that Merleau-Ponty's integration of space and time goes considerably further than Einstein's. The space-time of Einstein adheres to the classical formula whereas the phenomenological "time-space" of Heidegger and Merleau-Ponty surpasses it.

We may contrast Einsteinian and phenomenological approaches with respect to the here-now origin of space and time. Einsteinian physics is geared to move forward from the here-now, to move out of and away from the origin in such a way that 0,0 is taken as but the dimensionless point of departure for the objective measurements and analyses carried out by the detached subject. In this posture, the old subject-object dichotomy is unquestioningly maintained. What Merleau-Ponty achieves in his analysis of the "time-like" and "space-like" aspects of perspectival experience is, in effect, a *phenomenological reversal* of the Einsteinian orientation. "Shifting gears," Merleau-Ponty moves *backward* into the 0,0 origin, whereupon he finds that it is no mere transfer point in an external exchange between a detached subject and its object, but is the whole dimension of their internal relatedness. In so moving backward, the originating point "expands" or "thickens" into the action sphere of dialectical process. In sum, whereas Einstein moves away from the indivisible action sphere thereby reducing it to a dimensionless point in the space-time continuum, Merleau-Ponty's phenomenological reversal brings an awareness of the fullness of this primordial dimension from which space-time itself derives. But let me be more specific about just what the reversal discloses.

For Merleau-Ponty, the present, the here-now, possesses a "dialectical thickness" not found in Einstein's account. Again, in "space-like" perspectival experience, "The enigma consists in the fact that I see things, each one in its place, precisely because they eclipse one another, and that they are rivals before my sight precisely because each one is in its own place" (1964, p. 180). This paradoxical interplay of the visible and invisible suggests that what is eclipsed is not simply elsewhere (see Fig. 3.1); it also is here. Although the presencing of elsewhere here defies the Einsteinian mandate that they be taken as externally related points, this is not to say that here and elsewhere collapse into a single point of immediate presence. Constituting neither a dimensionless point nor points that merely are outside one another, here and elsewhere enact an internally mediated exchange that confers thickness on both. *Both*? Do we have two distinct terms, as in Einstein's analysis? Yes. But unlike what we find in Einstein, the two also are one!

We are led to a similar conclusion with respect to the "time-like" aspect of perspectival experience examined in Merleau-Ponty's earlier (1945/1962) account. That work was based on Heidegger's interpretation of temporality, and we have seen that, for Heidegger, true time is no succession of dimensionless now-points but possesses "three-dimensional thickness." To be sure, the future is withheld from the present, and the past is denied to the present. Yet both nevertheless make their presence felt in the thickness of the now. Thus, owing to the mutual interpenetration of past, present, and future, dialectical thickness is not only found in the here, but in the now as well. Consequently, the future is not simply *after* the present, nor is the past simply *before* it — the proscriptions of the Einstein-Minkowski light cones to the contrary notwithstanding. What we have instead is the internal dialectic in which future and past — each in its own way — presences now, without merely collapsing into that now.

Note that it is when here and now serve merely as points of departure for measuring space-like and time-like intervals that spatial and temporal relations themselves are dualistically divided. In this account, the light cones stretch out from the point of origin in such a way that the absolute future and absolute past are not only unambiguously set apart from one another, but from the elsewhere region of space-time as well. In the contrasting phenomenological approach — because elsewhere presences *here* and otherwhen presences *now* — elsewhere and otherwhen also are present to *one another*, each in its own way. We might imagine the light cones being retracted, reeled back into the here-now, yet not just deflating to a singular point at the center. In the paradoxical "thickening" of the here-now that results, space and time themselves overlap paradoxically. So the dialectic of here and elsewhere, of now and otherwhen, entails a dialectic of space and time themselves (see Abram 1996, Chapter 6). In the interchange, Merleau-Ponty's depth dimension and Heidegger's "time-space" are inseparably linked. The phenomenological philosopher Sue Cataldi (1993) confirms this in her study of Merleau-Pontean depth when she asserts, "Merleau-Ponty continues to associate depth...with the thickness of a present holding a past and a future together" (p. 67).

The dialectical exchange in question is of course no "objective event," given that the subject also abides in the dynamic thickness of here-elsewhere/now-otherwhen. To repeat Merleau-Ponty's way of putting it, the primal dimension of depth "goes toward things from, as starting point, this body to which I myself am fastened" (1964, p. 173). In thus grounding the "I" in the body, the "I" ceases to be a high-flying Cartesian *cogito* poised only for the forward thrust. A *backward* movement is involved in recognizing the "starting point" as densely embodied. A reversal takes place in which there is no longer a simple division between the subject's here-now and the things that lie elsewhere and otherwhen. Instead, an *internal circulation* of subject and object is set in motion, for the now-embodied subject is "one of the visibles" (Merleau-Ponty 1968, p. 135). In this regard, it is worth repeating what Merleau-Ponty said about the experience of artists like Cézanne. In precise contrast to Einsteinian experience,

The painter's vision is not a view upon the *outside*, a merely "physical-optical" relation with the world. The world no longer stands before him through representation; rather, it is the painter to whom the things of the world give birth by a sort of concentration or coming-to-itself of the visible. Ultimately the painting relates to nothing at all among experienced things unless it is first of all "autofigurative"....The spectacle is first of all a spectacle of itself before it is a spectacle of something outside of it. (1964, p. 181)

In short, the phenomenological reversal of the classical posture enacted by Merleau-Ponty and Heidegger allows us to apprehend the "starting point" or origin of our action as a whole dimension of lived embodiment and dialectical process.

The inability of contemporary physics to achieve unification was explored in the previous chapter. The obstacle is discontinuity. In addressing the challenge of quantum gravity, a confrontation with the radical discontinuities of sub-Planckian action and black holes is inescapable. It is here that hope is dashed of imposing unity upon nature by containing it within the space-time continuum — despite all the valiant efforts of string theory and other ingenious strategies that seek to calm "the violent quantum jitters" (Greene 1999, p. 158). In switching from space-time to the time-space of Heidegger and Merleau-Ponty, a kind of unity can now be achieved. The catch, of course, is that it is not the idealized unity to which the *cogito* has long aspired. Instead Heidegger and Merleau-Ponty offer us a paradoxical unity-in-diversity, or continuity-in-discontinuity. To be sure, this is a difficult offer for us to accept. In our long-term pursuit of idealized continuity, we have sought only to suppress discontinuity. The fear has been that if we were to accept fundamental discontinuity we would succumb to the chaos of the *apeiron*. I suggest, however, that if we could actually bring ourselves to such acceptance in the manner proposed by phenomenology, instead of continuity simply dissolving and discontinuity prevailing one-sidedly, we would find ourselves enacting a *dialectic* of continuity and discontinuity in which the continuous aspect would be stronger than ever, since it would no longer be posited speciously so as to conceal the underlying discontinuity. Let me digress a little in order to clarify this dialectic.

*

In his critique of Jean-Paul Sartre's *Being and Nothingness* (1943/1956), Merleau-Ponty comments that Sartre applies the ancient formula, *"being is, nothingness is not"* (Merleau-Ponty 1968, p. 88); in so doing, Sartre presupposes the classical division of these two terms. Being is *en soi*; it is the pure positivity of the object. In stark categorial opposition to this is the *pour soi*, the sheer negativity of the subject. "It is with this intuition of Being as absolute plenitude and absolute positivity, and with a view of nothingness purified of all the being we mix into it, that Sartre expects to account for our primordial access to the things" (p. 52). However, says Merleau-Ponty, "from the moment that I conceive of myself as negativity and the world as positivity, there is no longer any interaction" (p. 52). In his attempt to achieve interaction, Sartre does paradoxically juxtapose being and nothingness, but the two do not blend organically and the outcome is but an ambivalent oscillation between mutually contradictory poles that "belies the inherence of being in nothingness and nothingness in being" (p. 73). Whichever pole we consider, whether "the void of nothingness or the absolute fullness of being," what we ignore are "density, depth, the plurality of planes, the background worlds" (p. 68).

For his part, Merleau-Ponty insists that being is not pure positivity, that negativity is native to it, is incorporated within its very core. Thus, neither being nor nothingness exist in unalloyed immediacy; instead they mediate each other internally and the result is *depth*:

> The negations, the perspective deformations, the possibilities, which I have learned to consider as extrinsic denominations, I must now reintegrate into Being — which therefore is staggered out in depth, conceals itself at the same time that it discloses itself....If we succeed in describing the access to the things themselves, it will only be through this opacity and this depth, which never cease: there is no thing fully observable, no inspection of the thing that would be without gaps and that would be total....Conversely, the imaginary is not an absolute unobservable: it finds in the body analogues of itself that incarnate it. (1968, p. 77)

The mutual mediation of being and nothingness reflected in the experience of depth leads Merleau-Ponty to conclude that "for the intuition of being and the negintuition of nothingness must be substituted a *dialectic*" (1968, p. 89). According to Merleau-Ponty:

> Dialectical thought is that which admits reciprocal actions or inter-actions....Dialectical thought is that which admits that each term is itself only by proceeding toward the opposed term, becomes what it is through the movement, that it is one and the same thing for each to pass into the other or to become itself, to leave itself or to retire into itself....Each term is its own mediation....
>
> Is not the dialectic...in every case the reversal of relationships, their solidarity throughout the reversal, the intelligible movement which is not a sum of positions...but which distributes them over several planes, integrates them into a being in depth?....In sum...[dialectical thought is] capable of differentiating and of integrating into one sole universe the double or even multiple meanings, as Heraclitus has already shown us opposite directions coinciding in the circular movement....The circular movement is neither the simple sum of the opposed movements nor a third movement added to them, but their *common meaning*, the two component movements visible as one sole movement. (pp. 89–91)

Merleau-Ponty goes on to discuss the difficulty of maintaining a genuine dialectical approach, the persistent tendency for dialectical thinking to lose its edge and lapse back into classical positivity. He characterizes this "bad dialectic" as one that, in its anxiety "to get to the end" (p. 93), to reach a final closure, "results in a new positive, a new position" (p. 95). In this play on words, we see a connection between classical positivity and classical *space*, the classical *continuum*, which is indeed a "sum of positions," a static framework for juxta-position. Recalling what Merleau-Ponty says about Cartesian space in "Eye and Mind," we recognize its identity with what he subsequently says of Sartrean being: it entails "absolute positivity." Again, the continuum is a "space without hiding places which in each of its points is only what it is....Every point of space is and is thought to be right where it is — one here, another there" (1964, p. 173). Cartesian space, then, is "everywhere equal to itself, homogeneous" (1964, p. 173); it is a purely "positive being, outside all points of view, beyond all latency and all depth, having no true thickness" (1964, p. 174).

If there can be no "hiding places" in such a space, it goes without saying that there can be no gaps or holes in it. This renunciation of discontinuity that splits it off from continuity is but a variant expression of the general cleavage of nothingness and being within the classical mindframe. Given that, in this way of approaching space, there is no room whatever for the negative, for discontinuity, if discontinuity should nonetheless arise, it must be regarded as an unwanted incursion to be avoided, repressed, or eliminated.

We know, of course, that discontinuity *has* arisen on the contemporary scientific scene, that, indeed, it is a primary characteristic of contemporary science. In the previous chapter, I gave the Hilbert-space treatment of microphysical discontinuity as an example of how, under the continuing influence of classical thought, the effort has been made to avoid discontinuity, and how this actually has lead to the *perpetuation* of discontinuity on an implicit level. In its essence, the Hilbert-space stratagem is an attempt to circumvent discontinuity by a proliferation of continua that, in point of fact, are discontinuous with respect to each other.

In his critique of Sartre, Merleau-Ponty helps us to understand just why, when we divide continuity and discontinuity in an absolute way, all of our efforts to avoid the latter are in vain. With Sartre, says Merleau-Ponty, "we are beyond monism and dualism, because dualism has been pushed so far that the opposites, no longer in competition, are at rest the one against the other, coextensive with one another....compound and union are impossible between what is [i.e., being, continuity] and what is not [nothingness, discontinuity], but, for the same reason that makes the compound impossible, the one could not be thought without the other" (1968, pp. 54–55). Continuing in the same vein, Merleau-Ponty states: "as absolutely opposed, Being and Nothingness are indiscernible. It is the absolute inexistence of Nothingness that makes it need Being....It is precisely because Being and Nothingness, the yes and the no, cannot be blended together...that, when we see being, nothingness is immediately there" (p. 64). And finally: "I said in turn that 'nothingness is not' and 'being is' are the same thought — and that nothingness and being are not united. Connect the two: they are not united precisely because they are the same thing in two contradictories = ambivalence" (footnote on p. 69). In other words, classical thinking, when carried to its logical conclusion, eventuates in an absolute differentiation of being and nothingness, of continuity and discontinuity, that amounts to their *lack* of differentiation, their indistinguishability. It is for this reason that the one-sided insistence upon absolute continuity necessarily must bring absolute discontinuity, as we have seen in the examples cited above. What the idealized classical formulation cannot give us is the *dialectical integration* of continuity and discontinuity, the blending of these in depth that constitutes the reality of the lifeworld.

It is true that, until the nineteenth century, scientific analysis could proceed successfully by taking the classical idealization of space as the reality; for all intents and purposes, discontinuity could be ignored. But then, in carrying its program toward completion in the late nineteenth and twentieth centuries, science came to be confronted with the fact that the repression of discontinuity is no longer possible. We have seen the ineffectiveness of contemporary attempts to preserve continuity under these circumstances. Thus, it seems that, since we can no longer convincingly deny discontinuity, we have little choice but to accept it. Of course, we would gain nothing by accepting discontinuity while continuing to operate within the still-prevalent paradigm of dualism, since that would only mean passively resigning ourselves to the absolute nullification of space. The constructive acceptance of discontinuity requires that we come to understand it *dialectically*, to grasp it in terms of the lifeworld dimension of depth that supersedes its categorical separation from continuity.

However, while the dialectical approach does offer the promise that science may be able honestly to face up to discontinuity without simply abandoning its quest for unity, much work remains to be done by way of fleshing out the dialectic. If the forces of nature are truly to be unified in a dialectical way, phenomenological dimensionality will need to be delivered in concrete detail, not just in broad philosophical outline. To begin this process, we turn to *topology*.

Note

1. In this and succeeding chapters, it should be helpful to bear in mind the difference discussed in the preface between the term "phenomenology" as it is commonly employed in science and its philosophical usage in this book.

Chapter 4
Topological Phenomenology

4.1 Introduction

In string theory, the vibrational interactions of strings, one-dimensional subatomic filaments, reflect the fundamental properties of matter and the operation of nature's basic forces (in M-theory, a recent elaboration on string theory, higher-dimensional versions of strings called "branes" enter the picture). The challenge to string theorists is to solve their general equations in such a way that the vibrational patterns are described in fine detail, and unequivocally. If this aim could be achieved, the long sought unified account of nature would be realized. But I noted in Chapter 2 that, while many in the scientific community are excited over the prospect that string theory can succeed at this, the theory has not yet reached a mature stage of development. A significant obstacle is the high degree of abstraction to which string theory has had to resort in its effort to attain unification within the old paradigm of object-in-space-before-subject. (See Chapter 2 for a discussion of other, related problems associated with string theory.)

In the less abstract physics of the Newtonian era, the gap between the mathematical representation of physical reality and dynamic reality itself could be kept to a minimum. In the classical equations of motion, for example, changes observed in nature were directly expressed. This capacity to bring concrete change and formal description into close alignment was indeed of critical importance to the program of classical physics. The theoretical account of nature was validated by making it strictly answerable to the observed facts of nature. Theory and fact could be aligned in this way because they made use of the same order of space and time. Theoretical space and time were no different from that serving as the intuitive framework for immediate observation.

By contrast, modern physics — commencing with Einstein and continuing with quantum mechanics, Kaluza-Klein theory, the standard model of unification, and string theory — permits a measure of contact to be lost between form and fact. They are removed from each other by an act of abstraction in which the spatiotemporal framework is transformed. We saw in Chapter 2 that, in Kaluza-Klein theory, the space and time of immediate experience are no longer taken as composing the epistemological framework for establishing the invariance or symmetry of object relationships; instead they are now regarded as themselves objects within a more abstract, higher-dimensional epistemological framework unknowable by direct intuition. It is because the "primordial continuum" of Kaluza-Klein theory upon which string theory relies does not constitute the directly intuited framework for our experiences that there can be no direct correspondence between theory and observation. But let me be more specific about the problem posed for string theory by the opening of this gap.

String theory adopts the Kaluza-Klein account of cosmogony introduced in Chapter 2. On this view, submicroscopic strings embodying all four forms of physical interaction are conceived as originally vibrating in a ten-dimensional universe that was perfectly symmetric and scaled around the Planck length (in Chapter 8, we will consider the eleven-dimensional version of string theory given by M-theory). Then a "dimensional bifurcation" took place in which a subset of dimensions expanded relative to the remaining dimensions, giving the 3+1-dimensional visible universe known to us today. The six dimensions that did not expand are hidden from view, being curled up at the Planck length. Coincident with dimensional bifurcation, the initially perfect symmetry of vibrating strings that had encompassed the four forces of nature began dissolving into the four separate and seemingly unrelated force-field symmetries observable in the contemporary universe. (In the standard model, the specific physical mechanism proposed for non-gravitational symmetry breaking events is the interaction of the symmetric force fields with fields of the Higgs type that are assumed to have formed with the progressive cooling of the early universe. A more detailed picture of cosmogony will be provided in Chapters 9 and 10.) It is the compactness of the six-dimensional sub-manifold, its ultra-microscopic concealment, that creates the mistaken impression that the universe consists solely of 3+1-dimensional space-time; mathematical descriptions of the forces of nature have been so different from each other because they have been written within that limited dimensional context, thereby neglecting the hidden symmetry that arises within the broader dimensional context. The essential task of unification is seen as recovering the hidden symmetry.

Lifting the veil on the hidden symmetry requires more than mathematically demonstrating that strings vibrate in six dimensions beyond the ones visible in the macroscopic universe. The precise shape of those dimensions must be specified if the theory is to be brought into contact with the facts of physical reality. Implicit in this conventional approach to dimensionality is its *objectification*. The 3+1-dimensional and six-dimensional submanifolds in which force particles are purported to vibrate are objectified topological spaces to be described within a ten-dimensional analytical continuum or epistemological space. However, since the latter is not given to us intuitively, as was the analytical space of Newtonian physics, the topological objects embedded in ten-dimensional space in principle cannot be directly related to immediate experience and thus must necessarily be determined by an indirect, less-convincingly grounded procedure. So it is not only that "we would require an accelerator the size of the whole universe" in order to directly observe the topological environment of submicroscopic strings (Greene 1999, p. 215). The problem of direct observation is insoluble when operating in the objectifying mode because — unlike the situation in classical physics — the dimensionality of the objects to be observed exceeds the 3+1-dimensional framework of the observer.

The methodological difficulty of theoretically unifying the forces of nature without the aid of laboratory observation derives specifically from the fact that the particular vibrational patterns of the force-bearing strings do depend on the exact topological form in which the hidden dimensions are compacted. If there were no latitude for variation in the topological parameters of the vibrational symmetry components, the way to recover the hidden symmetry would be clear. The fact, however, is that the number of possible topological solutions that can satisfy the equations of string theory is enormous. It is not that the theory provides *no* guideline for choosing among the many alternatives. String theory does lay down certain requirements for the geometric setup of the six hidden dimensions and research has determined that a class of topological structures known as Calabi-Yau spaces can meet these requirements (Candelas, Horowitz,

Strominger, and Witten 1985). But hardly does this narrow the field of possibilities sufficiently, since, within the restricted Calabi-Yau class, a huge array of possibilities remains. Generally, we are faced with the fact that string theory — entailing as it does a higher order of abstraction than that aligned with concretely observable reality — provides no satisfactory means for directly ascertaining the precise six-dimensional topological structure presumed to be concealed at the Planck length. Therefore, when this approach is employed, efforts to recover hidden symmetry are cumbersome and work must proceed in the dark.

My critique of string theory's abstractness, and of the abstractness of modern physics in general, reflects a key issue in the foundations of mathematical physics: that of formalism vs. intuitionism. Intuitionism is rooted in the thinking of Descartes. "By intuition," says Descartes,

> I mean, not the wavering assurance of the senses, or the deceitful judgment of a misconstructing imagination, but a conception, formed by unclouded mental attention, so easy and distinct as to leave no room for doubt in regard to the thing we are understanding. It comes to the same thing if we say: It is an indubitable conception formed by an unclouded mind; one that originates solely from the light of reason, and is more certain even than deduction, because it is simpler….Thus, anybody can see by mental intuition that he himself exists, that he thinks, that a triangle is bounded by just three lines, and a globe by a single surface, and so on. (Descartes 1628/1954, p. 155)

This mode of insight was carried forward by Kant in his *a priori* certitude about space and time as the immutable organizing principle for all experience, and, early in the twentieth century, it was further advanced by certain neo-Kantian mathematical intuitionists (such as Brouwer and Weyl; see Kline 1980). We saw in Chapter 2, however, that the classical intuition of space and time already had been opened to serious doubt before the twentieth century began. As a consequence of this, it was the formalist approach to mathematical physics that came to prominence and has been dominant ever since. Here mathematical abstraction is employed in an effort to fill the vacuum left by the collapse of classical intuition. Beginning with Einstein's introduction of the counterintuitive notion of four dimensional space-time, which responded to the failure of classical space implicit in the Michelson-Morley experiment, physics has heaped abstraction upon abstraction, becoming more and more detached from the reality of everyday experience. In the Copenhagen Interpretation of quantum mechanics, for example, no bones are made about the fact that its Hilbert space is but a mathematical fiction devised to yield "pragmatic results" in the absence of an ability to represent the microphysical domain in an intuitively grounded, realistic manner. And while opponents of the Copenhagen view have argued for a "realistic" construal of the quantum domain, the "reality" most describe arises as a mere artifact of the *formalism*.

This is true of what physicist Henry Stapp calls the "absolute psi interpretation" (1979, p. 12). "Psi" refers here to the basic mathematical wave function of quantum mechanics, described by a vector in Hilbert space. In the "absolute psi interpretation" (which includes the view that the psi wave collapses upon observation and the alternative, "many worlds" view that it does not), the mathematical wave function is seen as a veridical representation of the underlying quantum reality. But I would ask whether a "reality" of this kind, being naught but a byproduct of the formalism devised by the analyst, can be regarded as anything more than *nominally* real. It is in essentially the same sense that string theory's Kaluza-Klein account of unification is "realistic" (as Witten [1981] claims). I suggest that the subsymmetries of Kaluza-Klein do not reflect concrete reality as much as the need to uphold abstract continuity. Although most physicists have been willing to ignore the limitations of such a formalistic approach, the minority group seeking a form of quantum mechanical realism that is more than merely nominal remains sensitive to the problem (e.g., Stapp 1979, Bohm 1980, Josephson 1987). However, this does not mean that formalism should be abandoned in favor of *traditional* realism or intuitionism.

The mere renunciation of formalism in advocacy of classical intuition cannot address the problem of unification. I submit that the basic difficulty with the abstractionist approach to unification lies not in its questioning of classical intuition but in its failure to question it deeply enough. Though contemporary string theory implies that unification indeed cannot be achieved if we restrict ourselves to the Kantian intuition of space and time, instead of raising genuine doubt about that intuition by opening continuity itself to question — which would in fact call into question the classical formula of object-in-space-before-subject — we have seen that string theory maintains the continuity abstraction in a higher-order form. Thus, both the need to question classical intuition and the intention of upholding it are implicitly expressed. By indicating that the old intuition of space and time should not be taken *a priori* after all, that it actually resulted from a concrete cosmogonic process, the question is raised; by at once assuming that the continuous space and time of immediate experience are but topological manifestations in a more abstract order of changeless continuity, the question is cancelled. Having obtained an idea of the price the abstractionist pays for cancelling his own question before it earnestly can be asked, we may ponder an alternative.

Suppose the question is not cancelled. Suppose a *genuine* concretization of mathematical physics is entertained. Instead of sidestepping the loss of Kantian symmetry, of epistemological continuity, by rendering the 3+1-dimensional continuum a mere topological object in a more abstract continuum, suppose we hold our ground at the first level of abstraction and *allow* the classical intuition of space and time to be challenged. Could mathematical physics, which has always given primacy to invariance or symmetry, survive such a development?

To question the principle of symmetry is of course to venture beyond it. But rather than merely forsaking the symmetry of space and of the object relations therein for some notion of nonsymmetry that is foreign to it, what if nonsymmetry could be given *internal* expression? Then space would be dynamized, recognized as changing from within itself. Might not this concretely self-transforming space provide an intrinsic directive for the details of cosmogony? For one thing, in the absence of a gap between levels of abstraction, there would be no free-floating topological objects requiring indirect empirical determination. Moreover, in a *full-fledged* concretization of mathematical physics, the gap would be closed between the objects-in-space and the observing subject. With the analyst no longer detached from the objects of analysis, all structures would appear to him or her as immediately intuited concomitants of the transformation process, giving a full and natural unification of the forms of interaction manifest in nature — a *unity-in-diversity*, that is, since the aspect of nonsymmetry and change would be incorporated in such a way that the old, one-sided, idealized unity would be surpassed.

4.2 Phenomenological Intuition, Topology, and the Klein Bottle

From its inception in the seventeenth century, modern science has been governed by the root intuition of object-in-space-before-subject. What we see in string theory is a last ditch, essentially unsuccessful effort to maintain that intuition in the face of microphysical phenomena that are fundamentally incompatible with it. We know from our work in the previous chapter that an alternative does exist. The intuition of objects in space cast before the fixed gaze of a detached subject can be supplanted by a *phenomenological* intuition of the dialectical interplay of object, space, and subject. In the concrete transpermeation of these terms that would then take place, continuity and discontinuity, symmetry and asymmetry, and changelessness and change would be dynamically integrated.

However, if phenomenological dimensionality, and, in particular, Merleau-Ponty's depth dimension, are to be brought to bear on the problem of unification now confronting physics, phenomenological intuition must be expressed with greater clarity and precision. This is where topology enters the picture, although, to be sure, I am now no longer speaking of the abstract, objectified approach to topology currently employed in mathematical physics.

In conventional mathematics, topology is generally defined as the sub-discipline that concerns itself with those properties of geometric figures that remain invariant when said figures are stretched or deformed (without tearing). The etymology of the word provides a better indication of topology's inherent concreteness compared with other branches of mathematics. Topology is the study of *topos*, "place." The concrete character of this term is evidenced by its relationship to words like "posture": the root meaning of "posture" is "to place." Philosopher of science John Schumacher thus defines "posture" as "the way a thing makes a place in the world" (1989, pp. 17–18). For her part, the philosopher Maxine Sheets-Johnstone is able to demonstrate that, whereas Euclidean geometry involves practices that are largely disembodied, "topology...is rooted in the body" (1990, p. 42). Topology then seems to be an ideal discipline for the task at hand: to flesh out phenomenological intuition so that it may be applied to the problem of dialectically integrating the forces of nature.

Both Heidegger and Merleau-Ponty adumbrated the importance of topology for the phenomenological enterprise. In his meditation on the poetic character of thinking, Heidegger intimated that "poetry that thinks is in truth the topology of Being. This topology tells Being the where-abouts of its actual presence" (1954/1971, p. 12). Merleau-Ponty's ac-knowledgement of topology came in a preliminary working note for *The Visible and the Invisible* written in October 1959. Here he instructed him-self as follows:

> Take topological space as a model of being. The Euclidean space is the model for [idealized] perspectival being [and is consistent]...with the classical ontology....The topological space, on the contrary, [is] a milieu in which are circumscribed relations of proximity, of envelop-ment, etc. [and] is the image of a being that...is at the same time older than everything and "of the first day" (Hegel)....[Topological space] is encountered not only at the level of the physical world, but again it is constitutive of life, and finally it founds the *wild* principle of Logos — — It is this wild or brute being that intervenes at all levels to over-come the problems of the classical ontology. (1968, pp. 210–11)

Unmistakably, this topological space — involving as it does the rela-tions of "proximity" and "envelopment," and being "older than everything and 'of the first day'" — is equivalent to the concept of dimension Mer-leau-Ponty had outlined earlier (in "Eye and Mind"): the concept of *depth*. Can we sharpen our focus on the depth dimension by going further with topology? A curious topological structure appears especially promising in this regard: the *Klein bottle*.

An ordinary bottle conforms to conventional intuition regarding inside and outside. It is a container whose interior region is clearly set off from what lies outside of it. If we fill such a bottle with liquid, for example, and seal its cap, the fluid will remain enclosed — unless the surface is broken, in which case it will pour out. Although conventional containers are thus either open or closed, let us try to imagine a vessel that is *both*. I am not merely referring to a container that is *partially* closed (such as a bottle without its cap), but to a vessel that is completely closed and completely open *at the same time*. The liquid contents of such a strange vessel would be well sealed within it, and yet, paradoxically, they would freely spill out! The Klein bottle (Fig. 4.1) is a container of this sort. Its paradoxical structure flagrantly defies the classical intuition of containment that compels us to think in either/or terms (closed or open, inside or outside, etc.).

Figure 4.1. The Klein bottle (from Gardner 1979, p. 151)

The topological property of the Klein bottle that is responsible for its peculiar nature is its *one-sidedness*. More commonplace topological figures such as the sphere and the doughnut-shaped torus are two-sided; their opposing sides can be identified in a straightforward, unambiguous fashion. Therefore, they meet the classical expectation of being closed structures, structures whose interior regions *remain* interior. In the contrasting case of the Klein bottle, inside and outside are freely reversible. Let me try to shed more light on just what this means.

Elsewhere, I have used the Klein bottle to address a variety of philosophical issues (see Rosen 1994, 1995, 1997, 2004, 2006). For our present purpose, we begin with a simple illustration.

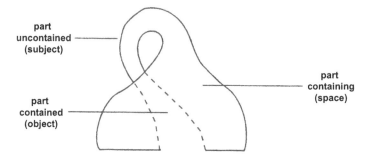

Figure 4.2. Parts of the Klein bottle (after Ryan 1993, p. 98)

Figure 4.2 is my adaptation of communication theorist Paul Ryan's (1993, p. 98) linear schemata for the Klein bottle. According to Ryan, the three basic features of the Klein bottle are "part contained," "part uncontained," and "part containing." Here we see how the part contained opens out (at the bottom of the figure) to form the perimeter of the container, and how this, in turn, passes over into the uncontained aspect (in the upper portion of Fig. 4.2). The three parts of this structure thus flow into one another in a continuous, self-containing movement that flies in the face of the classical trichotomy of contained, containing, and uncontained — symbolically, of object, space, and subject. But we can also see an aspect of *dis*continuity in the diagram. At the juncture where the part uncontained passes into the part contained, the structure must intersect itself (compare this with the corresponding feature of Fig. 4.1). Would this not break the figure open, rendering it *simply* discontinuous? While this is indeed the case for a Klein bottle conceived as an object in ordinary space, the true Klein bottle actually enacts a *dialectic* of continuity and discontinuity, as will become clear in the course of further exploring this peculiar structure. We can say then that, in its highly schematic way, the one-dimensional diagram lays out symbolically the basic terms involved in the "continuously discontinuous" dialectic of depth. Depicted here is the process by which the three-dimensional object of the lifeworld, in the act of containing itself, is transformed into the subject. This blueprint for phenomenological interrelatedness gives us a graphic indication of how the mutually exclusive categories of classical thought are surpassed by a threefold relation of mutual inclusion. It is this relation that is expressed in the primal dimension of depth. When Merleau-Ponty says that the "enigma [of depth] consists in the fact that I see things...precisely because they eclipse one another," that "their exteriority is known in their envelopment," he is saying, in effect, that the categorical division between the inside and outside of things is superseded in the depth dimension. Just this supersession is embodied by the one-sided Klein bottle. The Klein bottle therefore helps to convey something of the sense of dimensional depth that is lost to us when the fluid lifeworld relationships between inside and outside, closure and openness, continuity and discontinuity, are overshadowed in the Cartesian experience of their categorical separation.

However, must the self-containing one-sidedness of the Klein bottle be seen as involving the *spatial* container? Granting the Klein bottle's *symbolic* value, could we not view its inside-out flow from "part contained" to "part containing" merely as a characteristic of an object that itself is simply "inside" of space, with space continuing to play the classical role of that which contains without being contained? In other words, despite its suggestive quality, does the Klein bottle not lend itself to classical idealization as a mere object-in-space just as much as any other structure?

A well-known example of a one-sided topological structure that indeed can be treated as simply contained in three-dimensional space is the *Moebius strip*. Although its opposing sides do flow into each other, this is classically interpretable as but a global property of the surface, a feature that depends on the way in which the surface is enclosed in space but one that has no bearing on the closure of space as such; that is, the topological structure of the Moebius, the particular way its boundaries are formed (one end of the strip must be twisted before joining it to the other), can be seen as unrelated to the sheer boundedness constituted by the structureless point elements of space itself. So, despite the one-sidedness of the Moebius strip, the three-dimensional space in which it is embedded can be taken as retaining its simple closure. The maintenance of a strict distinction between the global properties of a topological structure and the local structurelessness of its spatial context is mathematics' way of upholding the underlying classical relation of object-in-space. Given that the Moebius strip does lend itself to drawing said categorical distinction, can we say the same of the Klein bottle? Although conventional mathematics answers this question in the affirmative, I will suggest the contrary.

Let us go beyond the linear schemata given in Figure 4.2 and obtain an idea of what the actual construction of a Klein bottle would entail.

Both rows of Figure 4.3 depict the progressive closing of a tubular surface that initially is open. In the upper row, the end circles of the tube are joined in the conventional way, brought together through the three-dimensional space outside the body of the tube to produce a torus. By contrast, the end circles in the lower row are superimposed from *inside* the body of the tube, an operation requiring the tube to pass *through* itself. This results in the formation of the Klein bottle. But in three-dimensional space, no structure can penetrate itself without cutting a hole in its surface, an act that would render the model topologically imperfect (simply discontinuous). So the

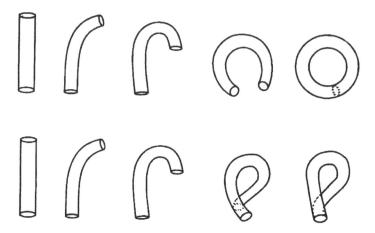

Figure 4.3. Construction of torus (upper row) and Klein bottle (lower row)

construction of a Klein bottle cannot effectively be carried out when one is limited to three dimensions.

Mathematicians observe that a form that penetrates itself in a given number of dimensions can be produced without cutting a hole if an *added* dimension is available. The point is imaginatively illustrated by mathematician Rudolf Rucker (1977). He asks us to picture a species of "Flatlanders" attempting to assemble a Moebius strip, which is a lower-dimensional analogue of the Klein bottle. Rucker shows that, since the reality of these creatures would be limited to *two* dimensions, when they would try to make an actual model of the Moebius, they would be forced to cut a hole in it. Of course, no such problem with Moebius construction arises for us human beings, who have full access to three external dimensions. It is the making of the Klein bottle that is problematic for us, requiring as it would a *fourth* dimension. Try as we might we find no fourth dimension in which to execute this operation.

However, in contemporary mathematics, the fact that we cannot assemble a proper model of the Klein bottle in three-dimensional space is not seen as an obstacle. The modern mathematician does not limit him- or herself to the concrete reality of space but feels free to invoke any number of higher dimensions. Notice though, that in summoning into being these extra dimensions, the mathematician is extrapolating from the known three-dimensionality of the concrete world. This procedure of dimensional proliferation is an act of abstraction that presupposes that the nature of dimensionality itself is left unchanged. In the case of the Klein bottle, the "fourth dimension" required to complete its formation remains an *extensive continuum*, though this "higher space" is acknowledged as but a formal construct; the Klein bottle per se is regarded as an abstract mathematical object simply contained in this hyperspace (whereas the sphere, torus, and Moebius strip are relatively concrete mathematical objects, since tangibly perceptible models of them may be successfully fashioned in three dimensions). We see here how the conventional analysis of the Klein bottle unswervingly adheres to the classical formulation of object-in-space. Moreover, whether a mathematical object must be approached through hyperdimensional abstraction or it is concretizable, the mathematician's attention is always directed outward toward an object, toward that which is cast before his or her subjectivity. This is the aspect of the classical stance that takes subjectivity as the detached position from which all objects are viewed (or, better perhaps, from which all is viewed as object); here, never is subjectivity *as such* opened to view. Thus the posture of contemporary mathematics is faithfully aligned with that of Plato and Descartes in whatever topic it may be addressing. Always, there is the mathematical object (a geometric form or algebraic function), the space in which the object is contained, and the seldom-acknowledged uncontained subjectivity of the mathematician who is carrying out the analysis.

Now, in his phenomenological study of topology, the mathematician Stephen Barr advised that we should not be intimidated by the "higher mathematician....We must not be put off because he is interested only in the higher abstractions: we have an equal right to be interested in the tangible" (1964, p. 20). The tangible fact about the Klein bottle that is glossed over in the higher abstractions of modern mathematics is its *hole*. Because the standard approach has always presupposed extensive continuity, it cannot come to terms with the inherent *dis*continuity of the Klein bottle created by its self-intersection. Therefore, all too quickly, "higher" mathematics circumvents this concrete hole by an act of abstraction in which the Klein bottle is treated as a properly closed object embedded in a hyperdimensional continuum. Also implicit in the mainstream approach is the detached subjectivity of the mathematician before whom the object is cast.

I suggest that, by staying *with* the hole, we may bring into question the classical intuition of object-in-space-before-subject.

Let us look more closely at the hole in the Klein bottle. This loss in continuity is *necessary*. One certainly could make a hole in the Moebius strip, torus, or any other object in three-dimensional space, but such discontinuities would not be necessary inasmuch as these objects could be properly assembled in space *without* rupturing them. It is clear that whether such objects are cut open or left intact, the closure of the space containing them will not be brought into question; in rendering these objects discontinuous, we do not affect the assumption that the space in which they are embedded is simply continuous. With the Klein bottle it is different. Its discontinuity does speak to the supposed continuity of three-dimensional space itself, for the necessity of the hole in the bottle indicates that space is unable to contain the bottle the way ordinary objects appear containable. We know that if the Kleinian "object" is properly to be closed, assembled *without* a hole, an "added dimension" is required. Thus, for the Klein bottle to be accommodated, the three-dimensional continuum must in some way be opened up, its continuity opened to challenge. Of course, we could attempt to sidestep the challenge, to skip over the hole by a continuity-maintaining act of abstraction, as in the standard mathematical analysis of the Klein bottle. Assuming we do not employ this stratagem, what conclusion are we led to regarding the "higher" dimension that is required for the completion of the Klein bottle? If it is not an extensive continuum, what sort of dimension is it? I suggest that it is none other than the dimension of *depth* adumbrated by Merleau-Ponty.

Depth is not a "higher" dimension or an "extra" dimension; it is not a fourth dimension that transcends classical three-dimensionality. Rather — as the "*first* dimension" (Merleau-Ponty 1964, p. 180), depth constitutes the dynamic *source* of the Cartesian dimensions, their "natal space and matrix" (p. 176). Therefore, in realizing *depth*, we do not move away from classical experience but move back into its ground where we can gain a sense of the primordial process that first gives rise to it. The depth dimension does not complete the Klein bottle by adding anything to it. Rather, the Klein bottle reaches completion when we cease viewing it as an object-in-space and recognize it as the embodiment of depth. It is the Kleinian pattern of action (as schematically laid out in Figure 4.2) that expresses the in-depth relations among object, space, and subject from which the Kantian trichotomy is abstracted as an idealization. So it turns out that, far from the Klein bottle requiring a classical dimension for its completion, it is classical dimensionality that is completed by the Klein bottle, since — in its capacity as the embodiment of depth — the Klein bottle exposes the hitherto concealed ground of classical dimensionality. Here is the key to transforming our understanding of the Klein bottle so that we no longer view it as an imperfectly formed object in classical space but as the dynamic ground of that space: we must recognize that the hole in the bottle is a hole in classical space itself, a discontinuity that, when accepted dialectically instead of avoided (as discussed in Chapter 3), leads us beyond the concept of dimension as continuum to the idea of dimension as depth.

By way of summarizing the paradoxical features of the Klein bottle thus far considered, I refocus on the threefold disjunction implicit in the standard treatment of the Klein bottle: contained object, containing space, uncontained subject. To reiterate: (1) The contained constitutes the category of the bounded or finite, that of the immanent contents we reflect upon, whatever they may be. These include empirical facts and their generalizations (which may be given in the form of equations, invariances, or symmetries). (2) The containing space is the contextual bounded*ness* serving as the means by which reflection occurs. (3) The uncontained or unbounded is the transcendent agent of reflection, namely, the subject. It is in adhering to this classical trichotomy that the Klein bottle is conventionally deemed a topological object embedded in "four-dimensional space." But the actual nature of the Klein bottle suggests otherwise. The concrete necessity of its hole indicates that, in reality, this bottle is not a mere object, not simply enclosed in a continuum as can be assumed of ordinary objects, and not opened to the view of a subject that itself is detached, unviewed (uncontained). Instead of being contained in space, the Klein bottle may be said to *contain itself*, thereby superseding the dichotomy of container and contained. Instead of being reflected upon by a subject that itself remains out of reach, the self-containing Kleinian object may be said to *flow back into* the subject thereby disclosing — not a detached *cogito*, but the dimension of depth constituting the dialectical lifeworld.

4.3 The Physical Significance of the Klein Bottle

4.3.1 Spin

Having related Merleau-Ponty's depth dimension to quantum physical action in the previous chapter, and having now fleshed out that lifeworld dimension via Kleinian topology, the physical significance of the phenomenologically constituted Klein bottle is presently clear. The self-containing, spatio-sub-objective Klein bottle embodies $\hbar$, the quantized action associated with the emission of radiant energy in the microworld. In point of fact, this connection is already implicit in the standard formulation of subatomic spin, though the relationship is well disguised.

Subatomic particles are endowed with internal angular momentum or *spin*. According to the mathematical physicist Roger Penrose (1971), spin is the "most obvious physical concept that one has to start with, where quantum mechanics says something is discrete" (p. 151). The fundamental quantized action of spin, indexed by $\hbar/2$, hardly takes the form of a simply continuous spinning in three-dimensional space. When Wolfgang Pauli sought to model quantum mechanical spin, he employed the mathematics of complex numbers (involving the imaginary i), and, in particular, the "hypernumbers" developed by William Clifford in the nineteenth century (see Musès 1977, Applebaum 2000). Mathematician David Applebaum observes that Clifford's abstract algebraic research had been "motivated by geometry...particularly the problem of trying to generalize the properties of complex numbers so as to be able to describe rotations in three dimensions" (2000, p. 3). Pauli used the "simplest non-trivial example of a Clifford algebra" (p. 4) to derive three matrices, which yield the three components of electron spin:

$$S_x = \hbar/2 \, (0 \quad 1 \quad 1 \quad 0)$$
$$S_y = \hbar/2 \, (0 \quad -i \quad i \quad 0)$$
$$S_z = \hbar/2 \, (1 \quad 0 \quad 0 \quad -1),$$

where $\hbar/2$ is the basic unit of electron spin. In fact, $\hbar/2$ is taken as the basis for determining the spin of *all* subatomic particles. Fermions, which are typically matter particles like the electron and quark, have spin values that are odd-number multiples of $\hbar/2$ (such as $1\hbar/2$, $3\hbar/2$, $5\hbar/2$, etc.). Bosons, which are typically force particles like the photon or gluon (associated with electromagnetism and the strong nuclear force, respectively), entail spin values that are even-number multiples of $\hbar/2$ (such as $2\hbar/2$, $4\hbar/2$, etc.). In the standard approach, fermions conform to Fermi-Dirac particle distributions in which the Pauli exclusion principle is obeyed, whereas bosons conform to Bose-Einstein distributions in which the exclusion principle is not obeyed.

Note that, while the Pauli spin matrices employ the hypernumber i, they are based on a form of hypernumber that actually goes beyond i. The mathematician Charles Musès called this number ε, defined as $\varepsilon^2 = +1$, but $\varepsilon \neq \pm 1$. Musès associated each of Pauli's three matrices with a different variety of ε (1976, p. 213). But Musès was not satisfied with a merely algebraic expression of ε. Emphasizing the intimate relationship between algebra and geometry, he asserted that "geometry…can lead us to a deeper understanding" than the mere "brute facts" of algebra (1977, p. 77):

> The hypernumber i needs only a plane for visualization because, as it is multiplied by itself, it rotates in a plane as does a unit radius when a circle is drawn. But the hypernumber ε involves *reflection*….It is clear that in order to turn a right hand into its reflected version (a left hand), more is required than sliding or rotation in tri-dimensional space. It can be shown that a right hand would be changed into a left hand if it were rotated 180° *out* of our triple dimensional space into four-dimensional space, and then back into our space again….Therefore, because ε deals with reflections and not only with simple rotations as does i, its operations demand geometrically a four-dimensional space, whereas the simple, rotational character of i's operations demands only two-dimensional space. (p. 77)

Musès further observes that when the "kinematic geometry of epsilon" is applied to physics, it becomes "clear that [electron] spin is more like successive rotation through a four-dimensional space and then back into ours, rather than an ordinary, three-dimensional spinning. As of this writing, it is realized among physicists that such 'spin' is not ordinary; but ε clarifies what it is" (p. 77).

Elsewhere (Rosen 1994, p. 104), I related the kinematic geometry of ε to the topology of the Klein bottle. Like ε rotation, Kleinian action can also be seen as involving a "successive rotation through a four-dimensional space and then back into ours"; moreover, the Klein bottle possesses the property of reflection that transforms left into right and vice versa, as we will see in Chapter 6. I would maintain, of course, that the "extra dimension" into which Kleinian action flows is in fact no objectified fourth dimension but the sub-objective dimension of *depth*. This is the crucial distinction between the proposed phenomenological approach to the foundations of physics and the still-prevalent Cartesian one.

It might not be too much to say that all microworld dynamics arise from spin of the Kleinian kind ($\varepsilon\hbar/2$). The fundamental role of subatomic spin is brought out by the physicists F. A. M. Frescura and Basil Hiley (1980). These theorists associate spin with "the pregeometric structure of the holomovement" (p. 27) — the dynamic substrate that Bohm portrayed as underlying space, time, and quantum mechanics (see below). The significance of spin is also highlighted in an article by physicist David Hestenes entitled, "Quantum Mechanics from Self-Interaction" (1983). According to Hestenes, electromagnetic self-interaction is the core element of quantum mechanics, and spin is its "most characteristic feature" (p. 68). In emphasizing the central importance of spin, Hestenes indicates that the uncertainty relations ($\Delta x \Delta p_x = \hbar/2$) found in every aspect of quantum theory derive from spin: "We now see the uncertainty relations as consequences of a zero-point motion with a fixed zero-point angular momentum, the spin of the electron. This explains why the limiting constant $\hbar/2$ in the uncertainty relations is exactly equal to the magnitude of the electron spin" (p. 73).

Citing Hestenes and others, Huping Hu and Maoxin Wu (2004) venture considerably further in arguing for the fundamental role played by spin in physics and beyond. Their basic proposition is well summarized in the title of their article: "Spin as Primordial Self-Referential Process Driving Quantum Mechanics, Spacetime Dynamics and Consciousness." Hu and Wu thus do not limit themselves to the claim that spin is the source of physics. Primordial spin is "protopsychic" (p. 45) as well as protophysical; it is "the seat of consciousness and the linchpin between mind and brain" (p. 43). Or we may say that spin is psychophysical or sub-objective, not just an objective event taking place in physical space or a subjective happening enclosed within the psyche. Moreover, Hu and Wu, in portraying the self-referential nature of spin, come close to recognizing its Kleinian character. Associating spin with the particle self-interaction of which Hestenes speaks, they liken it to the recursive "strange loops" described by Douglas Hofstadter (1979). Here there is "an interaction between levels in which the top level reaches back down towards the bottom level influencing it, while at the same time being itself determined by the bottom level" (2004, p. 45). What we find when we turn to Hofstadter himself is that the "strange loop" may be related to the Klein bottle (1979, p. 691).

Evidently then, quantized microphysical process is rooted in Kleinian spatio-sub-objectivity. We may relate this primal action to the *apeiron*, the "wild or brute [order of] being" (Merleau-Ponty 1968, p. 211) in which subject, object, and space fluidly commingle (see Chapter 2). The connection with *apeiron* helps bring out another key feature of micro-dimensional process, its *vorticity*.

4.3.2 Vorticity

As should be clear by now, dimensional action is decidedly different from the classical form of action that takes place within a dimensional framework that is itself presumed to be *in*active. The latter kind of action is continuous; it consists of point-to-point mechanical motions through the continuum. But discontinuity is not simply absent from the classical scene. Rather, it makes its "presence" felt in the *negation* of continuity. And this is just what we have in the prime formulation of the classical stance: object-in-space-before-subject. Continuity lies on the side of object-in-space and discontinuity on the side of subject. A central problem for classical thinking is that, while object and subject are sharply divided from one another, somehow they must interact. This is of course the mind-body problem that has haunted traditional philosophy from its inception. We may readily identify the dualism here with the juxtaposition of contradictories that Merleau-Ponty found in Sartre, the "intuition of being and negintuition of nothingness" for which Merleau-Ponty sought to substitute "a *dialectic*" (1968, p. 89). Stated in terms of action, classical thinking juxtaposes continuous movement with its negation thereby creating an implicit contradiction.

It is different with the quantized spinning of the depth dimension. Although action here possesses an aspect of indivisibility that precludes it from being simply contained in the infinitely divisible continuum, this is not to say that such action is *just* discontinuous, discontinuous *as opposed to* continuous. Were we to say that, in effect we would merely be reducing lifeworld process to the "nothingness" of classical subjectivity. Instead, depth-dimensional action must be thought as the *dialectical blending* (not contradictory juxtaposition) of continuity and discontinuity. In this section, the dimensional dialectic will be clarified further by demonstrating that, when we look more closely at the cyclical spinning in question, its specific form proves to be *cyclonic*. Whereas simple cyclical action involves movement that merely is positive, i.e., movement from one delineable position to another, the action of a cyclone — or, more generally, of a vortex — blends "positive" and "negative," since it is a circulation through a center that is empty.

In ancient Greek cosmology, the vortex was employed as a fundamental model of natural process. Philosophers such as Anaximander, Anaximenes, Empedocles, Anaxagoras, and Democritus used the principle of vortical motion to explain the creation of the world. According to Anaximander, it is by the action of a vortex that all boundaries arise from the boundless archaic flux of the *apeiron*. Two millennia later, the idea of the vortex came back into prominence with Descartes, who postulated that the universe is a dynamic plenum filled with matter in vortical motion. But Isaac Newton questioned the plausibility of Descartes's concept and vortex theory was eclipsed once more until the middle of the nineteenth century. At that time, the physicist William Thomson (Lord Kelvin), taking his cue from Helmholtz's mathematical formulation of vortex motion, proposed that the atoms constituting all matter were none other than minute vortical disturbances in the ethereal sea then believed to pervade the whole of nature. However, the repudiation of the ether concept by the Michelson-Morley experiment (see Chapter 2) led once again to the discrediting of vortex theory, at least temporarily.

More recently, quantum theorists have entertained the idea of a "new ether." The quantum mechanical notion of the *vacuum* is relevant in this regard. Generally speaking, the vacuum or "ground state" is defined as the "stationary state of lowest energy of a...system of particles" (Lapedes 1978, p. 429). In quantum mechanics, the particles from which our universe is built are viewed as excitations of the undifferentiated vacuum or ground state. According to Bohm, the state in question is a "cosmic background of energy," an invisible, ether-like plenum serving as "the ground for the existence of everything, including ourselves" (1980, p. 192). For Bohm, this cosmic ground is "an unbroken and undivided totality," a "holomovement" that implicitly "enfolds" or "carries" all orders of manifestation, the way a "radio wave" carries or guides a "signal" (Bohm 1980, pp. 150–51). Speaking of how a vortex of water suddenly rises up from the ocean "as if from nowhere and out of nothing," Bohm speculated that "something like this could happen in the immense ocean of cosmic energy, creating a sudden wave pulse [signal] from which our 'universe' would be born" (1980, p. 192). Thus, according to Bohm, quantum mechanics (and relativity theory) can best be regarded as calling for a view of the world as an "Undivided Wholeness in Flowing Movement" (p. 11), a primary universal flux from which apparently stable forms of both matter and thought "form and dissolve...like ripples, waves, and vortices in a flowing stream. ...All matter is of this nature....In this flow, mind and matter are not separate substances. Rather, they are different aspects of one whole and unbroken movement" (p. 11).

Bohm's thinking has been carried forward by the physicist R. M. Kiehn. Bohm had proposed that the quantum mechanical wave function is guided by the ether-like cosmic ground via a "quantum potential" (see Bohm and Hiley 1975), a non-local wave field that pilots its corresponding particle not by imparting energy to it, but in the subtler fashion of providing it with information about the whole environment. Kiehn expressed this quantum potential in the specific terms of a vorticity distribution in a viscous compressible fluid: "The vorticity distribution replaces the Bohm quantum potential" (1999, p. 1). Commenting on Kiehn's work, Hu and Wu claim it indicates that "[vortical] spin is the process driving quantum mechanics" (2004, p. 43). Hu and Wu also take note of Penrose's long-term program for deriving space, time, and quantum mechanics from vortex-like algebraic structures known as "twistors," structures that Bohm and Hiley (1984) sought to generalize via "Clifford algebra as a possible basis for describing Bohm's 'implicate order'" (Hu and Wu 2004, p. 44). It seems then that contemporary physics may be inadvertently returning to Anaximander's *apeiron*, "the boundless giver of boundaries" (Rosen 2004, p. 135) that creates the world through the action of a vortex.

*

Proceeding intuitively, let us now look more closely at the structure of the vortex. Vortices sometimes appear to take the topological form of a torus, a structure that can be described as a ring-like circulation about an empty center. But the fact is that the torus does not provide us with the best model of natural cyclonic action. The "positive" and "negative" regions of the torus — its peripheral surface area or volume and its central hole — are strictly set apart from one another, for the hole in the torus is located *outside* the body of this simply closed structure. Generally speaking, this is not the way vortices actually manifest themselves in nature. In reality, the positive and negative portions of a whirlwind or whirlpool are blended dialectically in the fashion of the Klein bottle, whose hole is *internalized*, arising as it does from the act of self-containment.

In a book entitled *Sensitive Chaos*, Theodor Schwenk (1965) offers a detailed study of the manifold natural forms a vortex can take, and makes this claim about the manifestation of the vortex:

> It is a fundamental process — an archetypal form-gesture in all organic creation, human and animal, where in the wrinkling, folding, invaginating processes of gastrulation, organs for the development of consciousness are prepared. Forms arising out of this archetypal creative movement can be found everywhere in nature. (p. 41)

Schwenk's primary example of vortex generation is the forming and cresting of a water wave:

> If we could watch the process in slow motion we would see how a wave first rises above the general level of the water, how then the crest rushes on ahead of the surge, folds over and begins to curl under...forming hollow spaces in which air is imprisoned in the water....This presents us with a new formative principle: the wave folding over and finally curling under to form a circling vortex. (1965, p. 37)

Figure 4.4 is my adaptation of Schwenk's illustration of vortex generation. At the outset, we have a pool of water that is entirely flat and calm (Fig. 4.4(a)); there is no circulation of liquid in this initial state of affairs, and no airy, curved hollow space from which rotating water is excluded.

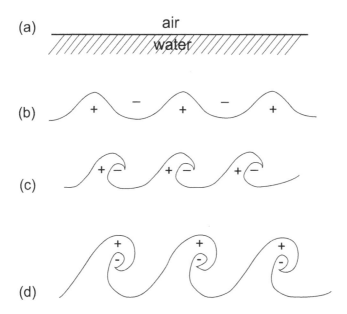

Figure 4.4. Stages in the formation of a vortex (adapted from Schwenk 1965, p. 37)

Let us say that, initially, the media of water and air are simply and completely partitioned from each other; they are juxtaposed without being opposed, so that, in effect, they are not dialectically engaged.

In the next stage of vortex generation, the water has begun to churn, resulting in the production of a wave train (Fig. 4.4(b)). We now clearly do have the beginnings of a dialectical interchange. The two media have begun to enter into each other, the circulating water swelling up into the air, the air pressing down into the water to create a hollow trough in it; in the process of thus overlapping one another, the dialectical opposition of "positive" (i.e. aqueous) and "negative" (gaseous) domains is brought into play.

Figure 4.4(c) shows the waves now starting to crest. In so doing, they begin to fold over upon themselves. With their in-curving crests becoming elongated, what had been semi-circular waveforms (Fig. 4.4(b)) are now more fully circular. Commensurate with this increase in the positive feature of the developing aqueous vortex, there is an increase in the negative: the semi-hollows of Figure 4.4(b) have begun to deepen, to form themselves into full-fledged holes. In quantitative terms, the passage from stage b to c brings an increase in the angular velocity of the water, accompanied by a proportional decrease in its angular inertia and spin radius. The conservation of angular momentum that thus attends the folding in of the cresting wave is reminiscent of a skater tucking in her arms as she spins: her velocity increases in proportion to the decrease in the radius of her rotation.

Figure 4.4(d) depicts the completion of vortex generation. Here both the swirling aqueous "plenum" and the gaseous "void" have been brought to full expression. And this full-fledged differentiation of "wholeness" and "hole-ness" is at once an *integration*. For the water does not merely spin *around* its empty, airy center in the fashion of a torus; it spins *into* it, is "sucked" into the vacuous hub in a whirling, "screw-like spiral" (Schwenk 1965, p. 45). Thus, even though water and air are well differentiated in the mature vortex, there is no simple line of demarcation that partitions these media into categorically separate regimes. Rather, in the self-intersecting, Klein bottle-like flow of the vortex, the "different media...flow together... the one [being] taken into the other so that the hollow space, like a vessel, is filled with [the medium] of a different quality" (Schwenk 1965, p. 40). Schwenk states, in general, that whenever different surfaces thus come into contact, "like the surface between water and air, these surfaces will become waved, overlap and finally curl round" (p. 37). A little later, he observes:

> This important phenomenon — the curling in of folds or layers to create a separate organ with a *life of its own* within the whole organism of the water — does actually occur in the forming of organic structures.... Like vortices, organs have their own life: they are distinct forms within the organism as a whole *and yet in constant flowing interplay with it*. (1965, p. 41; emphasis added)

Continuing in the same vein, Schwenk asserts that a vortex is "a form which has separated itself off from the general flow of the water; a self-contained region in the mass of the water, enclosed within itself and yet bound up with the whole" (p. 44). And again, a vortex is

> a separate entity within a streaming whole, just as an organ in an organism is an individual entity, yet closely integrated with the whole

through the flow of vital fluids. An organ is orientated in relation to the whole organism and also to the surrounding cosmos; yet it has its own rhythms and forms inner surfaces of its own. (p. 47)

In the foregoing statements, Schwenk clearly is emphasizing that vortex generation — seen as an archetypal process that bespeaks the generation of all organic forms — entails both differentiation and integration (said archetypal process should not be confused with Plato's static archetypal forms). The vortex is differentiated and integrated within itself, and differentiated from/integrated with its surrounding environment. It is also clear that such differentiation and integration do not arise separately in development, one following the other in linear sequence, but are inextricably interwoven aspects of the same underlying process.

Our chief interest, of course, is with *dimensional* vortex generation, with the cyclogenesis of the lifeworld. Therefore, were we to limit ourselves to dealing with the vortex as an *object in space*, we would not be addressing what primarily concerns us — regardless of whether we would be describing vortex generation in the strictly quantitative terms of inertia, velocity, and momentum, or speaking of it more qualitatively as involving the relationship between the presence (+) and absence (–) of a circulating form (water, a living organ, etc.). The production of vortices within a static spatial framework is a far cry from the quantized vortical action of the depth dimension itself. Again, micro-dimensional action is "*pre*spatial" (Heidegger 1962/1972, p. 14; emphasis added) or "pregeometric" (Frescura and Hiley 1980, p. 27); it is precisely the action that first opens up the framework within which objects appear for the scrutiny of the allegedly detached subject. (The opening up of the framework is inconceivable from the classical standpoint that *presupposes* the framework.) This means that depth-dimensional spin is neither itself objectifiable, nor is it an act of objectification carried out by a reflecting subject. Rather, it is the *prereflective differentiating/integrating* of object and subject — or, of the field of objectification (the space-time continuum or plenum) and the subjectivity that, in effect, constitutes a *hole* in the field. In accordance with our analogy, just as objectively given vortex generation entails a progressive increase in the differentiation/integration of the circulation of water and the airy hollow at the center of the circle, *dimensional* vortex generation involves an increasing differentiation/integration of object and subject that establishes prereflectively the possibility for reflective operations. More will be said about dimensional cyclogenesis when we consider its stage-wise course in Chapters 6 and 7.

It is clear by now that quantized microphysical action — as *dimensional* action (rather than action within a static dimensional framework) — is apeironic at bottom, incorporating as it does nature's inherent variability; its indeterminacy, discontinuity, and subject-object transpermeation. In extant quantum mechanical analysis however, these depth-dimensional roots of quantum spin are not fully recognized and accepted for what they are. Instead the attempt is made to treat the prereflective action as if it were but an object of reflection situated in simply continuous analytic space (to whatever degree that space is abstractly expressed). It is not so much that analysts are oblivious to the *apeiron* as that they feel the need to tame its "wildness." "Wild being" (Merleau-Ponty 1968, p. 211) is thus relegated to a "black box." We need not be concerned about the chaos the "box" contains as long as it *can* be contained. This is done by incorporating the underlying Kleinian spin ($\varepsilon\hbar/2$) into the theory as a constant value in accounts that maintain in probabilistic approximation the old aims of objectifying and controlling nature. In this way, the intrinsic dynamism of the world is cordoned off, functioning as an isolated negativity within an otherwise purely "positive" treatment of nature. Implicit here is a *dualism* of positive and negative, of the static and dynamic, continuity and discontinuity, symmetry and asymmetry — one that ultimately fails in the face of quantum gravity, where the effort to achieve unification in a well-grounded and consistent manner reaches an impasse.

With the phenomenological reversal of the customary approach to microphysical action, the quarantine on the concrete lifeworld is lifted and the dualism surpassed. In the previous chapter, I examined the transition from Einsteinian space-time to phenomenological "time-space." Here I spoke of "switching gears," of reversing the forward inclination of conventional thinking to move out of and away from the here-now origin of space-time. Through phenomenological intuition, we move backward into that 0,0 origin, whereupon we discover that the origin is no mere transfer point in an external exchange between a detached subject and its object, but is the whole dimension of their internal relatedness, the indivisible action sphere of dialectical process from which space-time itself derives. We are now able to recognize this action sphere as the Kleinian realm of primordial cyclonic spin (a view that is consistent with the proposals of Penrose [1960, 1967], Frescura and Hiley [1980], and others, to the effect that space-time derives from pre-spatiotemporal spin). The same phenomenological reversal we required for the point origin of space-time is required for the spinning point-particle. Reversing gears, we must move backward into the point so that its hitherto concealed spatio-sub-objective Kleinian source is fully disclosed. In this way, the "black box" is opened up and the *apeiron* is allowed to pour out. And yet, at the same time, the dialectic of paradox enacted through the Klein bottle is such that nature's dynamism is kept well-contained. By virtue of the latter aspect, we have continuity, symmetry, invariance; by dint of the former, we have discontinuity, asymmetry, and non-invariance. The members of these complementing pairs are not merely juxtaposed incommensurables, as in Sartre's still dualistic "bad dialectic"; instead they are blended in depth, as in the dialectic intimated by Merleau-Ponty.

To be sure, enacting such a phenomenological reversal would entail a truly radical departure from science's standard operating procedure. It would call for new priorities, a new posture, a new intuitive grasp of object and subject. Instead of maintaining their stance as detached subjects seeking to arrest nature via equations that objectify, physicists would need to accept the transpermeation of subject and object by becoming active participants in nature's dynamic process. Why should physicists be willing to undergo such a dramatic transformation? I submit it is because there is no other way for them to achieve their goal of bringing the basic forces of nature into harmony. I intend to demonstrate that while nature's unity-in-diversity cannot be attained under the old intuition of object-in-space-before-subject, this goal of individuation can indeed be realized by adopting the posture of dialectical phenomenology.

Chapter 5
The Dimensional Family of Topological Spinors

5.1 Generalization of Intuitive Topology

By way of advancing toward an approach that can encompass all the forces of nature, let us consider an intuitive generalization of the topological account given in the previous chapter.

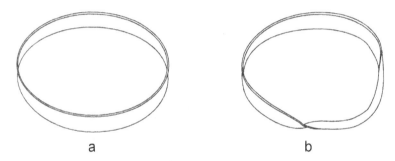

Figure 5.1. Cylindrical ring (a) and Moebius strip (b)

Mathematicians have investigated the transformations that result from bisecting topological surfaces (see Rosen 1994, 2004, 2006). To begin, we compare the bisection of the one-sided Moebius strip with that of its two-sided counterpart, the cylindrical ring. Suppose we were to cut the cylindrical ring of Figure 5.1(a) down the middle, proceeding along its full length. Upon completing the cut, the ring would simply decompose into a pair of identical narrower rings each possessing the same topological structure as the original. A more interesting result is obtained in bisecting the one-sided Moebius strip (Fig. 5.1(b)). Rather than falling into two separate pieces as one might expect, the cut surface retains its integrity but has now become the *two*-sided structure depicted in Figure 5.2.

Figure 5.2. The lemniscate

Compare a single side of the bisected Moebius strip with that of the two-sided cylindrical ring. Whereas revolution about the latter describes but one closed loop, traversal of the former gives us a doubly-looped, figure-8 pattern known in mathematics as a *lemniscate*; turned on its side, the lemniscatory surface resembles the familiar sign for infinity, ∞. The two patterns of movement are schematically contrasted in Figure 5.3.

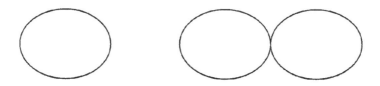

Figure 5.3. Schematic comparison of cylindrical (left) and lemniscatory (right) rotation

Now, the integrative quality of the Moebius surface lies in its paradoxical one-sidedness. The two sides of the Moebius flow unbrokenly into each other to form a single side, without either side actually losing its distinctness. When bisection of the Moebius strip transforms it into a *two*-sided structure, this integrity is lost. Yet we can now see that each of those sides, being lemniscatory in character, constitutes its own order of paradoxical "unity-in-diversity." That is, in a single side, we have a double cycle, these cycles being connected by a continuous movement through the central node of the figure-8. It is true that, in the lemniscate, we no longer have a complete *overlapping* of opposing elements, as we do in the Moebius. The lemniscate thus could be said to have less internal coherence than the Moebius. Be that as it may, a similar pattern of "transpolar flow" is evident in both structures.

Note that, whereas bisecting the simply symmetric cylindrical ring yields rings that are completely symmetric with respect to each other, bisecting the asymmetric Moebius strip produces lemniscatory sides that are related *enantiomorphically*; they complement each other in the manner of asymmetric mirror opposites. And just as the single lemniscatory side has its mirror counterpart on the other side of the bisected Moebius, the Moebius as a whole has its own enantiomorph. For while there exists but one form of the cylindrical ring, the Moebius surface can be produced in either a left- or right-handed version. If both versions of the Moebius were constructed, then "glued together," superimposed on one another point for point, the result would be a Klein bottle.

The Klein bottle, Moebius strip, and lemniscate constitute a series of topological forms that are nested within one another. Bisecting the Klein bottle produces Moebius enantiomorphs; bisecting the Moebius yields mirror-opposed lemniscatory sides. One more bisection is required to complete the series. Upon cutting the lemniscate, the surface neither retains its integrity nor simply falls into separate pieces. Instead, the single surface is transformed into two interlocking surfaces, each of which is itself lemniscatory (Fig. 5.4). The transformation brought about by this bisection is clearly the last one of any significance, since additional bisections — being bisections of *lemniscates*, can only produce the same results: interlocking lemniscates. The bisection series is completed then when we obtain interlocking lemniscates, a structure we shall henceforth refer to as the *sub-lemniscate*.

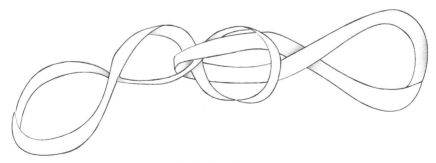

Figure 5.4. The sub-lemniscate

Now, to reiterate a key principle, conventional mathematics is guided by the classical intuition of object-in-space-before-subject. Accordingly, on the conventional view, all members of the bisection series are taken as two-dimensional topological objects embedded in a higher-dimensional continuum that is set up for the abstract operations of the detached mathematical analyst. What phenomenological thinking has told us is that the Klein bottle is no such object; instead, it is the circulatory action that *fuses* object and analyst, and, in the process, brings hitherto idealized three-dimensional space to its true dialectical realization. The *necessary hole* in the Klein bottle plays a critical role in this. We found in the last chapter that the bottle cannot be completed as an object in three-dimensional space without being ruptured. In phenomenologically exploring the "added dimension" required to fill this hole, we were able to confirm the idea that the Klein bottle — in its intimate blending of object, subject, and space — is an embodiment of the *depth* dimension, the three-dimensional lifeworld.

In considering the *sub*-Kleinian members of the bisection series, it is clear that they *can* be completed properly as objects in conventional three-dimensional space. It is for this reason that, while a paradoxical structure like the Moebius strip can well *symbolize* the integration of the three-dimensional subject and its object, it cannot effectively *embody* said integration. The proposition I now offer is that, although the sub-Kleinian members of our topological cyclonic spin family indeed cannot embody the dialectic of the *three*-dimensional lifeworld, they provide embodiments of *lower-dimensional* orders of depth.

Whereas the standard interpretation of the bisection series says that each of its members is a two-dimensional object embedded in a higher-dimensional continuum, what I am proposing is that, in depth-dimensional terms, *none* of its members actually possess this status (not even the Moebius, which *is* two-dimensional, but which — in its own lifeworld milieu — is no mere object). Since the necessary incompleteness of the Klein bottle in three-dimensional space is plain to us, it is relatively easy for us to see how this structure defies classical objectification. Why can we not see necessary holes in the sub-Kleinian members of the series? It is because our seeing is three-dimensional. Within this frame of observation, lower-dimensional dialectics are imperceptible. To understand the role of the sub-Kleinian structures vis-à-vis the lower dimensions, let us examine the exact manner in which lower-dimensional dialecticity is repressed within the three-dimensional framework.

In the foregoing chapter, we considered Merleau-Ponty's account of the dialectical opposition entailed in the perspectival interplay of the visible and invisible, wherein surfaces in space overlap and eclipse one another; all the surfaces of an object cannot be seen in a single glance. This is depicted in Figure 5.5(c); the diagram suggests the concealment of some of the cube's surfaces.

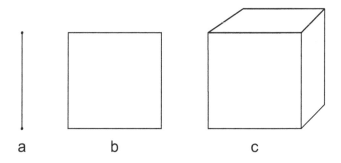

a   b   c

Figure 5.5. Endpoints of a line segment (a), lines bounding a surface area (b),
and planes bounding a volume of space (c)

But while the bounding planes that enclose a volume of three-dimensional space are not all simultaneously perceptible, the endpoints bounding the line segment (Fig. 5.5(a)) and the lines enclosing a surface area (Fig. 5.5(b)) *can* all be perceived at once. The simultaneous perceptibility of the points and lines renders these geometric beings purely and simply positive, as Merleau-Ponty said of Cartesian space in general. Though we may *speak* of them as being opposed, their appearance together before us as just what they are lacks dialectical tension. Therefore, while *planes* give evidence of the "invisible," and while, in so doing, they implicate our subjectivity (Merleau-Ponty 1968, p. 140), points and lines apparently do conform to the Cartesian idea of dimensionality. They appear as but the "in-itself" (Merleau-Ponty 1964, p. 173), as sheer positivity. Incapable of expressing true opposition, points and lines, in their simultaneous perceptibility, seem to give us only *juxta*position. It is true that standard mathematical analysis treats *all* dimensions, planes included, as merely juxtapositional continua. Phenomenological investigation tells us that such treatment is not appropriate when it comes to the interplay of planes in space. But are we not seeing now that classical analysis is indeed warranted in the case of points and lines?

I propose that conventional analysis does not suffice when we are inquiring into the deeper nature of lines and points. I suggest that while lines and points appear to be merely juxtapositional in the three-dimensional framework of observation, in actuality, they possess their own dialectical attributes. It is in the "repressed dialectics" of points and lines that we find the lower-dimensional lifeworlds that I have intimated.

To grasp the meaning of lower-dimensional dialecticity, let us begin by clarifying the classical distinction between three-dimensional space and the elements that bound it. Space is bounded by two-dimensional planes. Manifested in concrete form, the bounding plane serves to partition space into particular regions and neighborhoods. It is in its capacity as a concrete bounding element that the plane possesses opposing sides: the region of space on one side of the plane is set apart from that on the other. In the case of three-dimensional space itself, we cannot speak meaningfully of opposite sides since space does not function as a bounding element. Through its bounding elements, space plays the role that Plato attributed to it in the *Timaeus*: it is a *receptacle*, a medium or containing environment. For, to contain an object in space, one must establish a boundary that sets off its interior region from what lies outside of it. Note that while mathematicians often abstractly refer to the two-dimensional plane as a "space," in ordinary concrete experience it is only the three-dimensional domain that serves in this capacity. What of the line and the point in this milieu? They may be regarded as subsidiary bounding elements. The line partitions the planar bounding element, divides it into two-dimensional areas, and the point partitions the linear bounding element, cuts it into one-dimensional lengths. With this as a background, we may now consider the possibility of a lower-dimensional "receptacle," that is, a lower dimensionality that would actually function concretely as a space in its own right, not just as a bounding element for three-dimensional space.

Beginning with Edwin Abbott's imaginative excursion into "Flatland" (1884/1983), there have been many conjectures on what experience might be like for a lower-dimensional being. However, in much of this, the *dia-*

lectical character of dimensionality is missed, and this has resulted in some inconsistencies. Take, for example, the speculations of the Russian theosophical philosopher P. D. Ouspensky:

> Let us...consider the two-dimensional world, and the being living on a plane....In what manner will a being living on such a plane universe cognize his world? First of all we can affirm that he will not feel the plane upon which he lives. He will not do so because he will feel the objects, i.e., [the flat] figures which are on this plane. He will feel the lines which limit them....The lines will differ from the plane in that they produce sensations; therefore they exist. The plane does not produce sensations; therefore it does not exist. (1970, p. 53)

Thus, according to Ouspensky's extrapolation of classical space, just as we three-dimensional beings do not at all sense the space in which we are embedded but only the two-dimensional surfaces of the objects in this space, two-dimensional beings would be restricted to sensing the one-dimensional edges of the objects in their space.

Ouspensky further notes:

> Sensing only lines, the plane being will not sense them as we do. *First of all, he will see no angle.* It is extremely easy for us to verify this by experiment. If we hold before our eyes two matches, inclined one to the other in a horizontal plane, then we shall see one line. To see the angle we shall have to *look from above.* The two-dimensional being cannot look from above and therefore cannot see the angle. (1970, pp. 53–54)

These conclusions about two-dimensional experience appear to be directly contradicted by Ouspensky later in his text, when he speaks of two-dimensional beings as capable of perceiving *surfaces* (not just lines), of experiencing "simultaneously in two directions" (p. 89), of viewing *circles*, figures that possess angles of 360° (p. 92). Why the discrepancy? Why is it that, in one place, Ouspensky asserts that the perceptual capacity of a two-dimensional being is strictly one-dimensional whereas elsewhere he claims that it is two-dimensional? Perhaps Ouspensky's susceptibility to this error stems from the fact that, while his approach to dimension is classical, the truth of perception may be found in the non-classical realm that lies *between* dimensions.

Ouspensky's *modus operandi* is extrapolation by dimensional analogy. Thus, when he draws the conclusion that "the two-dimensional being has no idea of an angle" in the line it perceives (p. 54), analogical reasoning evidently should tell him that we three-dimensional beings would be just as unable to perceive an angle in the *surface*. It is clear that we are incapable of seeing a solid cube or sphere "simultaneously from all sides" (p. 87). Ouspensky makes the claim that we "can never see, even in the minute, any part of the outer world as it is, that is, *as we know it* [i.e., in three dimensions]. We can never see the desk or the wardrobe *all at once, from all sides and inside*" (p. 88). Evident in these two sentences is Ouspensky's presupposition of a peremptory division between three- and two-dimensional sense perception, one that leads to the conclusion that if we cannot see *all* sides of a solid object at once, then we are limited to viewing but a *single* side of it. According to Ouspensky, while we do *know* of the existence of the other sides of any solid object we encounter, this knowledge is not directly sensed; it derives from conceptual inferences we draw about the object. Is it true that our sense experience permits us to apprehend but a single surface of a solid at a time; that we cannot simultaneously sense a surface that is perpendicular to it; that we have no sense of the *angle* joining two or more surfaces of the solid? Obviously it is not. The phenomenological fact of three-dimensional perception is that it is partial: although we surely cannot view all six outer surfaces of a solid cube in simple simultaneity, we can see up to *three* of them in this way. Thus, we *can* view angles joining surfaces, though, to be sure, we cannot apprehend them as completely as we can the angles that connect the lines of a plane figure. These are the facts of perspective that Ouspensky glosses over in his classical approach to dimension. "We cannot imagine the cube...seen, not in perspective, but simultaneously from all sides," says Ouspensky (p. 87). In failing to qualify this further, Ouspensky allows the inaccurate impression that perspectival experience entails a *single*-sided view of the cube, one that would preclude the sensation of angle.

In Ouspensky's construal of dimension, the dialectical character of lived experience is neglected. At polar extremes — where we would experience either all sides of the cube in simple simultaneity or just a single side, but nothing in between — dialectical tension is absent. The vitality of the dialectic is found only by taking into account the intermediary zone, where there is *opposition*, not just simple juxtaposition. The *sine qua non* of the perspectival three-dimensional lifeworld is the opposition of the sides of surfaces. And it should follow by analogy that, in a *two*-dimensional lifespace, there would be a dialectic of *opposing edges of lines*. This is the lower-dimensional action that is occluded in three-dimensional space, where the edges of lines appear to be merely juxtapositional. Note that a lower order of *subjectivity* should be implicit in this action, since dialectical action is not merely objective but entails the interplay of object and subject. Thus, just as the dialectic of the plane in the three-dimensional lifeworld gives evidence of Merleau-Ponty's "invisible," and, in so doing, implicates our subjectivity, the dialectic of the line in the two-dimensional lifeworld should entail a subjectivity of its own.

If the Klein bottle embodies the dialectic of the three-dimensional lifeworld; and if there is in fact a second lifeworld, one of different dimensionality than the first; then I am proposing that this second dimensional order is embodied by the lower-dimensional counterpart of the Klein bottle, the structure that results from bisecting the Klein bottle, namely, the *Moebius* structure.

It is true that, in three-dimensional space, the Moebius appears as a well-formed topological object, a strip or surface that can readily be completed without the necessity of self-intersection. But remember what Rucker (1977) said about "Flatlanders" attempting to assemble a Moebius strip in a *two*-dimensional space: they would in fact be forced to cut a hole in it, just as we three-dimensional beings must do with the Klein bottle. So it seems that, in two-dimensional space, the Moebius structure indeed would have the special, self-intersective hole necessary for playing out the dialectic at that level. Said dialectic clearly would not involve the interplay of opposing sides of a surface, since there could be no such opposition in two-space. But note that, while a Moebius strip embedded in three-dimensional space does take the form of a one-sided *surface*, it also has the property of being one-*edged*.

Unlike the closed surfaces of the torus and Klein bottle, the cylindrical ring and the Moebius strip possess edges[1] (see Fig. 5.1). It is not only the opposing *sides* of the Moebius that flow into each other, but also its *edges*. To confirm this, run your index finger continuously along an edge of the Moebius until the whole length of the surface has been traversed. Upon returning to your point of departure you will discover that you have covered *both* edges of the surface. In contrast, tracing an edge of the cylindrical ring maintains the simple distinction between the edges.

Of course, in three-dimensional space, the opposition of edges lacks the dialectical force of opposing *sides*. Though we may describe the boundary lines constituting the edges of the Moebius as "opposed" to one another at any local cross-section of the strip, the fact that we can view them simultaneously (unlike locally opposed sides of the surface) places them in the same simply positive context, thus renders their relationship merely juxtapositional, not truly oppositional. And if there is no genuine perspectival opposition, the Moebius *integration* of edges must lack full-fledged *trans*perspectival dialecticity. But this would not be the case for the Moebius structure of "Flatland." In this *two*-dimensional space, the fusion of edges would possess the same dialectical character as the Kleinian fusion of perspectives in three-dimensional space.

In two-dimensional space, the Moebius structure would not be an open surface but a *line*, one that is paradoxically both open and closed, just as the Klein bottle is an open and closed surface. We are aware, however, that this Kleinian "surface" is no simple two-dimensional object in three-dimensional space but is a structure that brings three-dimensionality to its dialectical completion. By the same token, the Moebius "line" would bring *two*-dimensionality to its (w)holeness. To better appreciate the threefold distinction among the Moebius line, the Moebius surface, and the Klein bottle, let us consider the simpler case of three non-paradoxical counterparts: the circle, the cylindrical ring, and the torus (Figs. 5.6(a-c)).

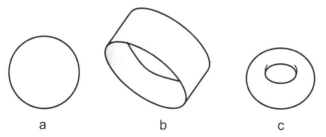

a b c

Figure 5.6. Circle (a), cylindrical ring (b), and torus (c)

We know that the cylindrical ring (Fig. 5.6(b)), taken as an object in three-dimensional space, is an open surface possessing two edges or bounding circles. This surface can be topologically transformed into a closed structure by elongating it, stretching it to form a tube, then bringing the circular edges of this tube together. The second part of this operation is identical to that depicted in the upper row of Figure 4.3. The closed surface that results is the torus (Fig. 5.6(c)). Now, if we proceed in the other direction, narrowing the cylindrical ring instead of elongating it, the circular edges draw closer and closer together. In the limit, we obtain but a single circle, the two-dimensional open surface having been reduced to a one-dimensional closed line (Fig. 5.6(a)).

Consider now parallel operations upon the Moebius counterpart of the cylindrical ring. Like the cylinder (Figs. 5.6(b) and 5.1(a)), the Moebius strip (Fig. 5.1(b)) is an open surface, one that locally possesses a pair of edges. If the Moebius were stretched in a manner similar to the elongation of the cylindrical ring, and if its edges were glued together, we would obtain the Klein bottle (Fig. 4.1). Of course, joining the edges of the Moebius strip is not so easy, given the odd one-edgedness of this structure (edges twist together to form a single edge). In fact, we are unable to execute the operation in three-dimensional space without tearing the surface, an action that topology does not permit. This limitation is familiar to us. What we are seeing is that the topological operation of identifying opposing edges of the Moebius strip is equivalent to that which we considered earlier, and which is shown in the lower row of Figure 4.3: the end circles of an elongated tube are joined from *inside* the tube's body, a procedure requiring the tube to pass through itself and thus to produce an impermissible breach when attempting to assemble the Klein bottle as an object in three-dimensional space. Moreover, identification of the opposing edges of a single Moebius strip is also equivalent to gluing together two Moebius strips of opposite orientation; we saw above that the latter operation yields the Klein bottle as well.

Now suppose, instead of elongating the Moebius strip, we made it narrower. What would happen in the limit of this operation? Would we obtain a simple circle as with the cylindrical ring, a simply closed line of a single dimension? To see what would actually result from this operation, let us compare the *perceptual* reduction of the cylindrical and Moebius strips in three-dimensional space.

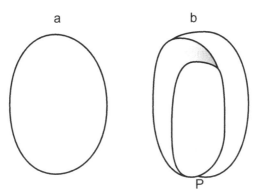

Figure 5.7. Edgewise views of cylindrical ring (a) and Moebius strip (b)

Figure 5.7(a) illustrates the fact that we three-dimensional observers can rotate a cylindrical ring in such a manner that only one of its edges is visible. In this way, the ring is perceptually reduced from a two-dimensional surface to a one-dimensional circle. It is clear from inspection of the Moebius that no such reduction is possible. The one-dimensional, one-edge view obtainable over the full length of the cylindrical ring can be realized in the Moebius case only at a *cross-section* of the strip. Note, moreover, that, in viewing the Moebius in this edgewise fashion, we do not actually see the cross-sectional line itself but just the endpoint of this line that is nearest to us, illustrated in Figure 5.7(b) by point P. However we position the Moebius, however we rotate it in three-dimensional space, at any given moment no more than a single point will be visible to us on the Moebius's edge at which extension in two dimensions will have vanished. It is only at this singular point that the perspectival opposition of sides is, in effect, perceptually neutralized (with no outer surface visible here, neither can we speak of an inner surface on the opposing side). Thus, the attempt to reduce the Moebius strip perceptually to a simple circle — a one-dimensional, purely positive presence in three-dimensional space — is frustrated by an unavoidable "dialectical surplus."

In the dimensional reduction thus far considered, only the properties of topological objects in three-dimensional space have been affected, not the characteristics of space itself, or those of the observer. Suppose, instead of merely reducing the dimension of our objects, we were able to change the whole dimensional frame of observation, lower the dimensionality of object-in-space-before-subject. We would then obtain the "Flat-land" state of affairs, that wherein space itself is two-dimensional. In this environment, the cylindrical surface would indeed become a circular line, but a line with peculiar properties when compared with our three-dimensional perception of lines. Just as we three-dimensional observers cannot fully view both sides of a surface at once but must experience them in perspectival opposition, the Flatland observer would not be able to view both edges of the circle at once. Contra Ouspensky, this would not mean that the Flatlander would simply be limited to perceiving but a *single* edge of the circle. Rather, the Flatlander would experience the edges of the circular line in a perspectivally oppositional fashion akin to the way we three-dimensional observers experience the sides of the toroidal surface. This dialectic of edges is certainly not detectable in the three-dimensional environment wherein lines are viewed in a strictly juxtapositional manner.

Now, in the foregoing dimensional reduction, the open-edged cylindrical surface is transformed into a circular Flatland line whose property of closure is akin to that of the torus, the higher-order counterpart of the cylinder in three-dimensional space. Similarly, the dimensional reduction of the open-edged Moebius band would yield a Flatland line whose structure would correspond to the band's higher-order counterpart, viz. the Klein bottle: unlike the open Moebius surface, the Moebius *line* would be both open and closed. Assuming then that the Flatland Moebius takes on the role played by the Klein bottle in three-dimensional space, would a Flatland structure exist whose role would correspond to that of the Moebius strip in three-space? The topological bisection series tells us that it would be the *one-dimensional lemniscate*.

Recall our study of the two-dimensional lemniscatory surface (Fig. 5.2). Though the opposing lobes of this dual structure are linked by a continuous movement through its central node, the lemniscate lacks the integrative quality of the one-sided Moebius band, whose opposing elements completely overlap one another. The dynamics of dimensional reduction suggest that, in Flatland, the lemniscate should lose its duality and assume the integrality of its higher-order Moebius counterpart. As the Moebius strip is a one-sided surface with an exposed edge, the Flatland lemniscate would be a one-edged line with an exposed *point*. And just as *closing* the exposed edge of the Moebius strip forms the Kleinian surface, the Flatlander would form the Moebius line by closing the exposed point of the linear lemniscate. Of course, in each milieu, a self-intersection would be required to complete the closure. We have seen that the necessary hole in the Klein bottle bespeaks its unobjectifiability in three-dimensional space. Following up on this observation led us to confirm the one-sided Klein bottle as the perspectivally integrative dialectical circulation of subject, object, and space constitutive of the three-dimensional lifeworld. A similar conclusion is called for with respect to the Moebius line of Flatland.

Above I noted the inability to eliminate the dialecticity of the Moebius surface in three-dimensional space (Fig. 5.7). The "dialectical surplus" of opposing Moebius sides in three dimensions betokens the intrinsic, *edge-wise* dialecticity of the Moebius in *two*-dimensional space: this one-edged line would be unobjectifiable in said space. The necessary gap in the Moebius line would intimate the dialectical spinning together of subject, object, and space constitutive of the two-dimensional lifeworld. To reiterate, the Moebius *surface* is indeed but an object in three-dimensional space, a structure that merely *symbolizes* the dialectic of three-dimensional depth; this dialectic can only be truly embodied via the Klein bottle. What I am suggesting, however, is that, when the Moebius is transposed into its own element, when it is given expression in the dimensional milieu of Flatland, which is the lifeworld of two-dimensional space and concomitant two-dimensional subjectivity, we then have the Moebius *line*, the unobjectifiable structure that fully embodies the dialectic of two-dimensional depth.

But we must take note of an essential *difference* between Kleinian and Moebial orders of dimensional depth. The Kleinian interplay of subject and object is more complex than its Moebius counterpart. In the three-dimensional Kleinian container, whatever the content of experience, all perception is framed by three degrees of dimensional differentiation, three kinds of bounding elements: planes, lines, and points (in subsequent chapters, we will examine the topo-phenomenological basis for generating these bounding elements). By contrast, Moebius experience would be limited to just two such terms: lines and points. This would seem to mean that the perception of objects in the "flat" Moebius lifeworld would be less differentiated, that the subject-object dialectic would be weaker in this environment. The perspectivity here, that of edges of the line, should be less sharply defined than the Kleinian perspectivity of sides that is familiar to us. Moreover, the dialecticity of lower-dimensional lifeworlds should become even weaker as we go further down.

The next lowest lifeworld order is that of the lemniscate (Table 5.1 summarizes the several topodimensional lifeworlds). In three-dimensional space, the lemniscate is objectively manifested via an open, two-sided surface (Fig. 5.2). Above we surmised that, in the *two*-dimensional milieu, the lemniscate would function as an open, one-edged line analogous to the open one-sidedness of the Moebius surface in three dimensions. A further dimension down, however, the lemniscate would enter its own element, the realm of one-dimensional space. Just as the open Moebius surface of three-dimensional space becomes the open-and-closed, self-intersective Moebius line of two dimensions, the open lemniscatory line of two-dimensional space would become an open-and-closed, self-intersecting *point* of a one-dimensional lifeworld (what Abbott might have called "Lineland"). If the Kleinian structure constitutes the dialectical circulation of subject and object by which the three-dimensional lifeworld is completed in earnest; and if the self-intersecting Moebius structure is the subject-object vortical spin cycle that completes the two-dimensional lifeworld; then the self-intersecting lemniscate would be the dialectical action that would bring to fruition the one-dimensional lifeworld. Still, with each step down to lower dimensionality, the dialectical tension of dimensional spin would be less, the perspectivity weaker; that is, the opposition and integration of the perspectives of objects, and of subject and object themselves, would be less sharply defined. The relative weakness of dialectical opposition in the lemniscatory lifeworld is indicated by the fact that here, only one element of differentiation would be operative, that of the point.

The last member of the bisection series is the sub-lemniscate. In three-dimensional space, this structure appears as the interlocking lemniscatory surfaces illustrated in Figure 5.4. How would the sub-lemniscate be manifested in the two-dimensional environment? Table 5.1 illustrates the general idea that, in moving into lower-dimensional lifeworlds, lower-order members of the bisection series assume the roles of their higher-order counterparts. Therefore, in the two-dimensional lifeworld, just as the Moebius structure would play the Klein bottle's role (it would be an open-and-closed integrative cycle or circulation), and as the lemniscate would play the role the Moebius played in three-dimensional space (it would be

TOPOLOGICAL ACTION CYCLE

LIFEWORLD DIMENSION	OPEN/CLOSED INTEGRATIVE CYCLE	OPEN INTEGRATIVE CYCLE	OPEN DUAL CYCLE	INTERLOCKING DUAL CYCLES
3d	Kleinian surface	Moebius surface	Lemniscatory surface	Sub-lemniscatory surfaces
2d	Moebius line	Lemniscatory line	Sub-lemniscatory line	
1d	Lemniscatory Point	Sub-lemniscatory point		
0d				

Table 5.1. Action cycles of the topodimensional lifeworlds

an open integrative circulation), the sub-lemniscate would function as the lemniscate did in three dimensions: it would be an open cycle of dual character, a two-edged line, with each edge tracing a lemniscatory pattern: ∞. Continuing the reduction, the sub-lemniscate of the one-dimensional lifeworld evidently would operate as an "open integrative point," thereby assuming the role that was played by the lemniscatory line in the two-dimensional lifeworld. Then, in the zero-dimensional realm, the sub-lemniscate would finally come into its own.

Dialecticity would be completely absent in the zero-dimensional sphere. This is borne out by a clear-cut difference between the zero-dimensional situation and those of higher dimensions. The three orders of lifeworld dialecticity are expressible in terms of the dimensional interplay of container, contained, and uncontained. To begin to appreciate the uniqueness of zero dimensionality, let us for the moment regard higher-dimensional containment relations in simplified conventional terms rather than concerning ourselves with the paradoxical *self*-containment that underlies them. We can then say that, at each level, the containment of objects is enacted by the lower-dimensional bounding elements of the containing space: the planes bounding three-dimensional space, the lines bounding two-dimensional space, and the points bounding one-dimensional space. These relations suggest the general idea that the $n-1$-dimensional bounding element of n-dimensional space itself serves as the containing space in relation to a lower-dimensional bounding element. For example, the two-dimensional plane that bounds objects in three-dimensional space would function as space itself in the lower-dimensional lifeworld wherein the highest-dimensional bounding element is the line.

What we encounter in making the transition to the zero-dimensional realm is the complete absence of bounding elements and thus, the absence of space, for there can be no spatial container if nothing can be contained. The zero-dimensional point that bounds or differentiates one-dimensional space does not translate as a spatial container possessing its own, lower-dimensional bounding element because there is no dimensional element below zero. This is reflected in the fact that — unlike the plane or the line — the point is utterly indivisible; it cannot be partitioned or subdivided in any way. It is clear that, given the absence of bounding elements in the zero-dimensional sphere, the dialectic of spatial container and objects contained cannot be enacted there.

There is, of course, a third term of the dialectic inseparable from the other two: the *un*contained, the subject. Obviously, if there is no spatial container and no contained object, neither could there be the detached vantage point of reflection upon the object that constitutes the subject. Since the zero-dimensional situation would be characterized by a complete lack of containment, there could be no *self*-containment here either, no lifeworld dialectic.[2] It is this total absence of dialectical structure that requires us to leave blank the zero-dimensional row of Table 5.1.

5.2 Topodimensional Spin Matrix

Now, Musès (1968) indicated a "higher epsilon-algebra" wherein "$\sqrt{\varepsilon_n}$ involves i_n, the subscripts of course referring to the $(n + 1)$th dimension since $i \equiv i_1$ already refers to D_2," (p. 42). Bearing in mind the intimate relationship between the Klein bottle and ε, we may adapt Musès's implication of a dimensional hierarchy of hypernumber values so as to express our geometric bisection series via algebraic notation. The three-dimensional Kleinian spinor can then be written ε_{D3}, with lower-dimensional members of the tightly knit spin family designated ε_{D2}, ε_{D1}, and ε_{D0} (corresponding to the Moebial, lemniscatory, and sub-lemniscatory circulations, respectively). The relationships among these values are given in Table 5.2, the topodimensional spin matrix.

ε_{D0}	$\varepsilon_{D0}/\varepsilon_{D1}$	$\varepsilon_{D0}/\varepsilon_{D2}$	$\varepsilon_{D0}/\varepsilon_{D3}$
$\varepsilon_{D1}/\varepsilon_{D0}$	ε_{D1}	$\varepsilon_{D1}/\varepsilon_{D2}$	$\varepsilon_{D1}/\varepsilon_{D3}$
$\varepsilon_{D2}/\varepsilon_{D0}$	$\varepsilon_{D2}/\varepsilon_{D1}$	ε_{D2}	$\varepsilon_{D2}/\varepsilon_{D3}$
$\varepsilon_{D3}/\varepsilon_{D0}$	$\varepsilon_{D3}/\varepsilon_{D1}$	$\varepsilon_{D3}/\varepsilon_{D2}$	ε_{D3}

Table 5.2. Interrelational matrix of topodimensional spin structures

In matrix representations, a square matrix like that given in Table 5.2 possesses a principal or main diagonal, the one extending from the upper left-hand corner to the lower right. It is here that the *eigenvalues*[3] of our topodimensional array are displayed.[4]

The word "eigenvalue" is a linguistic hybrid containing the German word *eigen*, meaning "own." "Eigenvalue" is a reflexive term indicating the primary elements or essential properties inherent in and unique to a given matrix: the "own values" or "self-values" of the matrix. The eigenvalues of a matrix have been alternatively expressed as the values that are "characteristic" of or "proper" to the matrix (an early definition of the word "proper" is "belonging to oneself or itself; own"[5]). In Heinz von Foerster's systems-theoretic application, "Eigenvalues, because of their self-defining (or self-generative) nature imply topological 'closure' ('circularity')" (1976, p. 93). My own reading of the eigenvalue is topo-*phenomenological*. An eigenvalue characterizes the self-referential, spatio-sub-objective action of a whole lifeworld.[6] The four eigenvalues of our topodimensional matrix thus correspond to our four basic orders of topological depth, whose interrelationships are specified two at a time in Table 5.2 by the matrix elements appearing off the principal diagonal.

To enhance our understanding of the Table, let us begin by considering the idea of the *stationary wave*. Such a wave is formed from the interference of two waves traveling in opposite directions. Figure 5.8 gives the four most basic configurations of stationary waves. These wave patterns can be created in objects such as musical instruments (by plucking a guitar string, for example). The waves here are quantized, divided into segments of equal size that are describable only in terms of whole numbers (we would not find a stationary wave with one and a half segments, for example). Figure 5.8(a) displays what, in music, is called the *fundamental tone* or first harmonic; this baseline frequency corresponds to the state of lowest vibrational energy. Figures 5.8(b) through 5.8(d) display the *overtones* of this ground state.

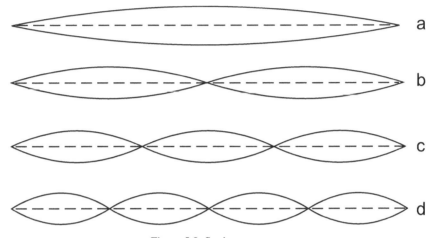

Figure 5.8. Stationary waves

Recall our Chapter 4 discussion of the ether-like microphysical ground state or vacuum that Bohm characterized as an "immense ocean of cosmic energy" (1980, p. 192). If the particles that make up our universe are excitations of this cosmic ground, we might imagine them as "overtones" of the ground, similar to the way Erwin Schroedinger (1961) initially conceived them. But Bohm's writing on this actually implies a subtler, fourfold distinction. First, there is the "cosmic background of energy" or oceanic "holomovement" serving as "the ground for the existence of everything, including ourselves" (1980, p. 192). From this universal flux, more differentiated forms of movement arise, such as radio waves or sound waves, which can carry signals or information that, in turn, can influence the behavior of matter and energy (Bohm 1980, pp. 150–51). The fourfold distinction then is among the holomovement, more differentiated forms of movement, the implicit information they carry, and the observable effects of this information in the physical world.

The topodimensional eigenvalues given in the Table 5.2 spin matrix correspond to four primary forms of movement. For now, we will focus on the relationships among these forms. The question of the information that is carried by the dimensional waves will be taken up in Chapter 10.

The Table 5.2 matrix meets the need for something more than the single harmonic series illustrated in Figure 5.8. For the overtones of such a series merely constitute higher registers of but one fundamental wave action, whereas each of the four orders of dimensional action we have been exploring must be seen as a fundamental action in its own right, a dimensional sub-objectivity constituting a unique eigenwave or self-identity (though lower-dimensional vortices, having lesser degrees of subject-object differentiation, have less sharply defined identities). The deeply qualitative differences among the four hypernumber identity operators is consistent with Musès's characterization of differing hypernumbers as "new forms of unity, which is to say new *kinds* of number, for as goes 1 [the identity operator], so go all numbers" (1968, p. 32). But it seems we should go even further. For, while the eigenvalues displayed on the principal diagonal of the matrix surely do not just differ quantitatively, neither do they merely express different qualities of the world; rather, as distinct spatio-sub-objectivities, they represent *different lifeworlds*. And yet they are intimately related.

What we have in the topodimensional wave matrix is not simply a single ground wave carrying information in the form of overtones, but a hierarchy of interrelated dimensional ground waves or "fundamental tones." Take the example of the two principal elements of highest dimensionality: ε_{D2} (the two-dimensional Moebius vortex) and ε_{D3} (the three-dimensional Kleinian vortex). In the matrix, these elements are linked by the combining elements given in the two corresponding non-principal cells, $\varepsilon_{D3}/\varepsilon_{D2}$ and $\varepsilon_{D2}/\varepsilon_{D3}$. Said coupling cells are related to each other *enantiomorphically*; they are mirror images of one another. The coupling cells signify the depth-dimensional counterparts of the conventionally observable, oppositely oriented Moebius strips which, when glued together, form the Klein bottle. Taken strictly as an eigenvalue, the depth-dimensional Moebius element is the open/closed, one-edged spin structure of the two-dimensional lifeworld. But when we shift our view of the Moebius, consider it in relation to higher, Kleinian dimensionality, a kind of "doubling" takes place in which the singular Moebius spin structure (ε_{D2}) becomes a pair of asymmetric, mirror opposed twins ($\varepsilon_{D3}/\varepsilon_{D2}$ and $\varepsilon_{D2}/\varepsilon_{D3}$). Whereas, in conventional observation, these are open Moebius surfaces in three-dimensional space, what we now require is the *depth-dimensional* realization of enantiomorphs.

Before proceeding, I must acknowledge a limitation of Table 5.2. As we are about to see explicitly, the dimensionalities given in the Table in fact are involved in developmental processes, events unfolding over time: the lower-dimensional fundamentals carry or support the self-development of the higher-dimensional ones. But the Table itself gives us only a synchronic view, with all matrix elements being displayed as simply co-present. In Chapter 7, the matrix will be "set in motion" to provide a more dynamic rendition of topodimensional generation (see Table 7.1). For the preliminary account presently given, we are obliged to read our synchronic table in a diachronic fashion.

When the Klein bottle is taken as an object in ordinary space, it is seen as simply deriving from the fusion of Moebius strip enantiomorphs. The depth-dimensional understanding of the Klein bottle suggests that this cannot be the whole truth. In view of the self-containing, self-transforming quality of the Klein bottle previously intimated, it appears we must regard this fundamental eigenstructure as developing at least partially from internal action, from dialectical action upon *itself*. What then is the underlying role of enantiomorphic action? Let us say that one enantiomorph, $\varepsilon_{D2}/\varepsilon_{D3}$, functions as a nascent form of the Kleinian wave engaged in dimensional self-generation, while the other, $\varepsilon_{D3}/\varepsilon_{D2}$, operates in the capacity of a Moebius carrier wave that guides Kleinian autopoiesis. Although each enantiomorph is a hybrid term spanning both dimensions, the $\varepsilon_{D2}/\varepsilon_{D3}$ enantiomorph uniquely expresses the role of the incipient Kleinian wave; its "fractionality" reflects the fact that it has not yet fully reached its potential as the integral three-dimensional structure, ε_{D3}. Whereas the "fractional" Kleinian enantiomorph falls short of its native dimensionality, the reciprocal $\varepsilon_{D3}/\varepsilon_{D2}$ Moebius enantiomorph *exceeds* its own two-dimensional sphere of action so it can function as a carrier wave of the higher-dimensional wave. The relation between enantiomorphs culminates with their *fusion*. We can say that, in completing their merger, enantiomorphs are "annihilated"; they cease to exist *as* free-standing entities. The Moebius enantiomorph, having fulfilled its role as carrier, is absorbed into the three-dimensional Kleinian vortex; the other enantiomorph — no longer a "fractional" dimensional hybrid, is brought to term as ε_{D3}, the fully three-dimensional Kleinian vortex itself.

To see how enantiomorphy in vortex generation can be observed in nature, we return to Schwenk (1965). In the course of his presentation, he depicts the generation of whole trains of vortices (Fig. 5.9).

a   b

Figure 5.9. Train of vortices (a) and rhythmic array of vortices (b)
(adapted from Schwenk 1965, pp. 38 and 51)

Figure 5.9(a) shows a progressive strengthening of vortical form in advancing from one vortex in the train to another. The figure highlights the enantiomorphy that is involved in cyclogenesis. We can see here how the clockwise action of a particular vortex in the train gives way to a counterclockwise action and how, from these two actions, there arises the next vortex in the train (clockwise and counterclockwise orientations are related enantiomorphically since mirror reflection changes clock sense). In commenting on Figure 5.9(b), Schwenk spells out the enantiomorphic aspect. Speaking of the "rhythmical arrangement of vortices" in a vortex train, he notes that "the vortices alternate in corresponding pairs, one slightly ahead spinning one way and the other, behind, spinning the other way" (1965, p. 51). Of course, when it comes to *dimensional* cyclogenesis, we are not dealing with objectively observable liquid flows continuously distributed in space. Instead vortical action is quantized and possesses a particulate aspect, allowing us to speak of the vortical Kleinian particle

($\varepsilon_{D2}/\varepsilon_{D3}$), whose self-generation is carried or catalyzed by the reciprocal Moebius particle ($\varepsilon_{D3}/\varepsilon_{D2}$) to which it is enantiomorphically related (the physical meaning of these particles will be examined in due course).

We see from the Table 5.2 matrix that the developing three-dimensional Kleinian wave culminating in ε_{D3} is not assisted exclusively by the two-dimensional Moebius wave, ε_{D2}, via the Moebius's $\varepsilon_{D3}/\varepsilon_{D2}$ enantiomorph, but by the two other vortices as well: the one-dimensional lemniscatory wave, ε_{D1}, and the zero-dimensional sub-lemniscatory wave, ε_{D0}. The linkage of ε_{D1} and ε_{D3} is given by the $\varepsilon_{D3}/\varepsilon_{D1}$ and $\varepsilon_{D1}/\varepsilon_{D3}$ coupling cells, and ε_{D0} and ε_{D3} are joined via $\varepsilon_{D3}/\varepsilon_{D0}$ and $\varepsilon_{D0}/\varepsilon_{D3}$. With respect to ε_{D1}, the Table discloses a second pair of enantiomorphs, $\varepsilon_{D2}/\varepsilon_{D1}$ and $\varepsilon_{D1}/\varepsilon_{D2}$. It is through this lower-dimensional coupling that the one-dimensional lemniscate guides the self-generation of the two-dimensional Moebius vortex. In terms of the series of topological structures observable to us in ordinary three-dimensional space, the lemniscate-Moebius relationship is expressed through the fusion of lemniscatory surfaces that yields the Moebius surface. Depth dimensionally, the lemniscate's support of the Moebius via $\varepsilon_{D2}/\varepsilon_{D1}$ is the lemniscate's *primary* guiding action, just as the $\varepsilon_{D3}/\varepsilon_{D2}$ enantiomorph of the Moebius provides primary support for the Kleinian wave. The second lemniscatory enantiomorph, $\varepsilon_{D3}/\varepsilon_{D1}$, constitutes a secondary source of guidance for the three-dimensional Kleinian vortex. Here the enantiomorphic liaison does not directly culminate in the production of the fully developed Klein wave. The fusion of $\varepsilon_{D3}/\varepsilon_{D1}$ and $\varepsilon_{D1}/\varepsilon_{D3}$ enantiomorphs leads instead to the $\varepsilon_{D3}/\varepsilon_{D2}$ and $\varepsilon_{D2}/\varepsilon_{D3}$ coupling, the higher-dimensional enantiomorphic pair whose subsequent merger does bring the Kleinian vortex to fruition.

We arrive at similar conclusions regarding the guiding action of the sub-lemniscatory wave, ε_{D0}. The sub-lemniscate possesses *three* enantiomorphic carriers. The first of these, $\varepsilon_{D1}/\varepsilon_{D0}$, affords support for the maturation of $\varepsilon_{D0}/\varepsilon_{D1}$ into ε_{D1} (the conventionally observable counterpart of the ε_{D0} and ε_{D1} relationship is given in the gluing together of sub-lemniscatory surfaces [Fig. 5.4] to yield the lemniscate [Fig. 5.2]). The second sub-lemniscatory enantiomorph, $\varepsilon_{D2}/\varepsilon_{D0}$, guides Moebius cyclogenesis in a secondary way, its fusion with $\varepsilon_{D0}/\varepsilon_{D2}$ leading to the $\varepsilon_{D2}/\varepsilon_{D1}$ and $\varepsilon_{D1}/\varepsilon_{D2}$ enantiomorphic relationship that climaxes in the self-realization of the two-dimensional Moebius vortex. Finally, the third ε_{D0} enantiomorph, $\varepsilon_{D3}/\varepsilon_{D0}$, supports Kleinian vortex generation in a tertiary manner: fusion of $\varepsilon_{D3}/\varepsilon_{D0}$ and $\varepsilon_{D0}/\varepsilon_{D3}$ → fusion of $\varepsilon_{D3}/\varepsilon_{D1}$ and $\varepsilon_{D1}/\varepsilon_{D3}$ → fusion of $\varepsilon_{D3}/\varepsilon_{D2}$ and $\varepsilon_{D2}/\varepsilon_{D3}$.

Regarding the Table 5.2 matrix in acoustical terms, the enantiomorphic pairings of elements suggest that "fundamental tones" of the matrix (the eigenvalues) are accompanied not only by "overtones" (the dimensional ratios with values >1, shown extending below the fundamentals) but also by mirror-image "undertones" (ratios with values <1, appearing to the right of the fundamentals). Harmonic theory does allow that, to an overtone series, a mirror-inverted undertone counterpart may be added.[7] Interestingly, what was likely the first theory of musical enantiomorphy was that given by Pythagoras, centered on the idea of the "music of the spheres." The basic design of the so-called *Pythagorean table* is displayed in Table 5.3.

1/1	1/2	1/3	1/4
2/1	2/2	2/3	2/4
3/1	3/2	3/3	3/4
4/1	4/2	4/3	4/4

Table 5.3. Section of the Pythagorean table

The Pythagorean table is usually portrayed as an indefinitely expanding series of musical intervals. What is shown in the limited section of the table that I have selected for illustrative purposes is a set of relationships that essentially corresponds to our topodimensional action matrix: There is a principal diagonal that contains a series of fundamental tones or self-values, these eigenvalues constituting "forms of unity" (Musès 1968, p. 32) that are taken as distinctive (1/1 ≠ 2/2 ≠ 3/3 ≠ 4/4). And the four principal intervals are coupled to each other two at a time by six enantiomorphically related pairs of overtone-undertone intervals. Perhaps then, we could go so far as to speculate that the values provided in Table 5.2 are the topodimensional counterparts of the Pythagorean musical intervals. These rhythmical relationships would then give us the "music of the dimensional spheres." I conclude this chapter with a relevant quote that prefigures our return (in Chapter 8) to the subject of *strings*:

> Music has long since provided the metaphors of choice for those puzzling over questions of cosmic concern. From the ancient Pythagorean "music of the spheres" to the "harmonies of nature" that have guided inquiry through the ages, we have collectively sought the song of nature in the gentle wanderings of celestial bodies and the riotous fulminations of subatomic particles. With the discovery of superstring theory, musical metaphors take on a startling reality, for the theory suggests that the microscopic landscape is suffused with tiny strings whose vibrational patterns orchestrate the evolution of the cosmos. (Greene 1999, p. 135)

Notes

1. Actually, the paradox of the Klein bottle is that it is both closed and *open*, whereas the edges of the cylindrical ring and Moebius surface make them simply open structures. The torus, for its part, is simply closed.

2. In *Topologies of the Flesh* (Rosen 2006), I explore in greater depth the philosophical implications of zero-dimensionality.

3. In linear algebra, an eigenvalue is the scale factor by which a vector (a directed magnitude) is multiplied in transformations that leave the vector otherwise unchanged. Such vectors are themselves called *eigenvectors*. Transformations involving eigenvectors and eigenvalues are elaborated through the use of matrices in this field of quantitative mathematical research. Though we also work with eigenvalues and their matrices in the present volume, a qualitatively different meaning of the term "eigenvalue" is brought out here.

4. Not all types of matrices have eigenvalues on their principal diagonals. A common example of a kind of matrix with eigenvalues that do appear here is the *diagonal matrix*. The Table 5.2 matrix does not seem to be a diagonal matrix, since all the entries of such matrices must be zero except for the ones appearing on the main diagonal. Yet we are going to see that the off-diagonal elements of our matrix in fact become "annihilated," leaving as the only values the terms on the main diagonal, which are the eigenvalues (see Table 7.1).

5. *The American College Dictionary*, Rodale ed., s.v. "proper."

6. It is interesting, perhaps, that, for Heidegger, the lifeworld opens from a prespatial dimension of action he termed *Ereignis*. Heidegger read this otherwise ordinary German word (meaning "event" or "occurrence") in a literalized fashion, understanding it as "making one's own," *Ereignis* being related to *eigen*, own. Heidegger thus

stated: "What determines both time and Being [i.e., the lifeworld], in their own, that is, in their belonging together, we shall call: *Ereignis*" (1962/1972, p. 19).

7. See Persichetti's discussion of musical "mirror writing" (1961, pp. 72–73) and Helmholtz's classical 1877 study of harmonic undertones.

Chapter 6
Basic Principles of Dimensional Transformation

6.1 Synsymmetry and the Self-Transformation of Space

Our ultimate concern is with the harmony of nature's fundamental forces. In this regard, I venture to suggest that each of the dimensional eigenwaves of Table 5.2 has its unique force field counterpart. More specifically, I propose that the ε_{D3} Kleinian wave is associated with electromagnetism, the ε_{D2} Moebius wave with the weak nuclear force, the ε_{D1} lemniscatory wave with the strong nuclear force, and the ε_{D0} sub-lemniscatory wave with gravitation. This proposition will gain credence with the concrete examination of the forces of nature that begins in Chapter 8. But first the ground will be prepared by considering in depth the principles of dimensional transformation that underlie the evolution of our universe from which the forces emerge. It is this task to which we turn in the present chapter and the next.

In closing the first section of Chapter 4, I intimated that dialectical unification of the forces of nature may be achieved by concretizing and dynamizing mathematical physics. The transformation anticipated would allow for the intrinsic interplay of symmetry and asymmetry, continuity and discontinuity, object and subject — instead of continuing to give priority to symmetry, continuity, and the division of subject and object, as is done under the old intuition. Such a radical change in approach requires a switch to phenomenological intuition. According to the prevailing view of cosmogony we considered earlier, all four forces of nature initially existed in a purely symmetric multi-dimensional space scaled around the Planck length. Subsequently, the perfect primordial symmetry was spontaneously broken by a dimensional bifurcation in which four of the original dimensions expanded to produce the visible universe we know today, with the other dimensions remaining hidden, thereby creating the appearance of irreconcilable differences among the forces. On this view, however, the "breaking" of symmetry actually entails no fundamental loss. While the original symmetry does become concealed, it is still there and the task of unification is merely to recover it. We have seen the limitations of this approach and know that the alternative phenomenological view of symmetry transformation involves a dialectic of symmetry and asymmetry — let us call it *syn*symmetry (Rosen 1975, 1994, 2004) — in which asymmetry is not relegated to a secondary position but plays an integral role. In the interest of clarifying this principle, it will be helpful to further explore the general nature of dimension.

We are seeking to bring to full expression a more dynamic intuition of space. We are attempting to better understand how dimensional transformation can be intuited as a concrete process of *self*-transformation (Chapter 4). But at the beginning of the twentieth century, when mathematical intuitionists were laying the topological foundation stones of contemporary dimension theory, the Kantian dictum of dimensional *invariance* was tacitly incorporated. To begin with, in 1911 Brouwer was able to prove that dimensions are not strictly homeomorphic with respect to each other: no operation is permissible by means of which one dimension can be put into 1:1 correspondence with another and, also, be continuously mapped into the other. What still was needed for a clear-cut demonstration of the topological invariance of dimension was the positive identification of a simple topological property responsible for the nonhomeomorphism. This came in 1912 with the intuitive reasoning of Poincaré. In Poincaré's scheme of classification, a given continuum could unambiguously be assigned a fixed integer value unique to it, its dimension number, said number depending on the kind of cuts required to divide it:

> Lines, which can be divided by cuts which are not continua [that is, by points], will be continua of one dimension; surfaces, which can be divided by continuous cuts of one dimension, will be continua of two dimensions; and finally space, which can be divided by continuous cuts of two dimensions, will be a continuum of three dimensions. (Quoted in Hurewicz and Wallman 1941, p. 3)

Subsequently, Poincaré's definition was precisely formulated by Brouwer and further refined by Urysohn and Menger (see Hurewicz and Wallman 1941, pp. 3–5). The effect of all this was to guarantee that dimensions themselves indeed are invariant, not subject to transformation. They were to be accepted ready-made, assumed to provide the uniform context within which all other geometric transformations were to be described.

It seems clear that, in Poincaré's intuition of the topologically invariant continuum, the "continuous cuts" are structureless boundary elements of dimension (points in a line, lines in a plane, etc.) that must retain their well-defined identities if they are to serve effectively in distinguishing one dimension from another in an unambiguous manner. If continuity were breached, boundary elements apparently would lose their sharp definition and topological invariance, the principle prohibiting the transformation of dimensionality, would be compromised. Then could the idea of self-transforming dimension not involve the transformation of the identity elements of dimension themselves? While mathematical physicists have not given much sustained attention to this possibility, a few relevant speculations have been offered.

In the context of asserting that "the concepts of spatial and temporal continuity are hardly adequate tools for dealing with the microphysical reality," Čapek (1961, p. 238) cited Menger's (1940) attempt to construct a "topology without points." Of course, it was not Menger's intention simply to drop the idea of space-time, so it was necessary for him to include *some* form of distinction. His dilemma led him to suggest a topology of "lumps" (instead of sharply defined points). Čapek quoted Menger's own misgivings in this regard:

> "By a lump, we mean something with a well defined boundary. But well defined boundaries are themselves the results of limiting processes....Thus, instead of lumps, we might use at the start something still more vague — something perhaps which has various degrees of density or at least admits a gradual transition to its complement. Such a theory might be of use for wave mechanics." (Menger, quoted in Čapek 1961, pp. 237–38)

The topology envisioned by Menger certainly runs counter to classical intuition: A "space" without categorically defined boundaries, one not possessing uniform infinite density, thus not infinitely divisible either, not reducible to structureless identity elements. A hint of the internally dynamic character of such an entity is given in Menger's suggestion that it should admit "a gradual transition to its complement."

Do we not already know of a structure possessing such characteristics? Is the inside-out Klein bottle not a prime exemplar of an entity with ambiguous boundaries, one that transforms itself dialectically so that it incorporates internal transitions to complementing aspects ("part contained" → "part containing" → "part uncontained"; see Fig. 4.2)? And, when non-classically grasped, does the Klein bottle not embody "various degrees of density"? For, the necessary hole in the Kleinian vortex constitutes a breach in the spatial continuum that reduces its density to zero. With the Kleinian circulation through its own hole, in fact we have an ongoing interplay of continuity and discontinuity in which density surely varies. Or we can say that the dynamic Kleinian structure is neither a whole possessing infinite density, a hole entailing zero density, nor merely a mixture of these in which the terms are externally related; rather, it is a thoroughly compounded *(w)hole*. Associating wholeness with symmetry and "holeness" with asymmetry, we can also see the intrinsically *syn*symmetric nature of Kleinian action. In flowing through itself as it does, the Klein bottle continually breaks and remakes its own symmetry, so to speak.

The phenomenologically construed Klein bottle well may be regarded as bringing out Menger's notion of an ambiguous boundary element of space. The elements of this self-transforming processual structure differ dramatically from the static, structureless point elements of the classical continuum. Moreover, in the paradoxical dimension of depth embodied by the Klein bottle, we cannot even strictly partition the "local" elements of space from space as whole (depth is "the experience of...a global 'locality'"; Merleau-Ponty 1964, p. 180). In the Kleinian deep, old notions of linearly scaled magnitude are confounded and "micro" and "macro" themselves interweave dialectically. (While the Klein bottle can indeed be thought of as a "global locality," its "local" aspect does involve bounding elements with characteristics that can be distinguished from those of the Klein bottle as a whole; see following chapters.)

Such an account of dimensional process surely "might be of use for wave mechanics," as Menger put it. However, what we are going to see in

Chapter 8 is that the Kleinian embodiment of quantum mechanical action accounts for only one force of nature, viz. electromagnetism. To incorporate the other three forces, the other three members of the topodimensional spin family must be brought into play. But we are not yet prepared to deal with this. For now, let us take another step toward clarifying the general nature of dimensional self-transformation.

6.2 From Symmetry Breaking to Dimensional Generation

The conventional description of cosmogony as a process of symmetry breaking carries with it an important implication. While talk of "breaking symmetry" might seem to suggest a negation of symmetry that is at odds with classical intuition, the very idea of symmetry *breaking* presupposes an initial state in which it was *un*broken. Presupposing a primordial state of perfect symmetry does put symmetry first, and one might wonder how a symmetry of this kind could be broken. In fact, some argue that the term "symmetry breaking" is actually a misnomer (see Liu 2002, p. 24), that the original symmetry is not really broken but only hidden (as I intimated above). The assumption of hidden symmetry is another way of maintaining classical intuition and giving primacy to changelessness. Alternatively we may say that underlying changelessness is mandated by supposing that dimensional transformations wrought by "symmetry breaking" are merely changes in the global topological structure of objectified space, not intrinsic transformations of the underlying analytical continuum as such.

In developing a new intuition of space, we cannot put symmetry first by speaking of "symmetry breaking" but must view symmetry itself as arising from a more primordial, *pre*symmetric condition, one in which symmetry is *already* "broken," as it were. Only when the nonsymmetric aspect is thus given a primary role (rather than being treated as a mere negation of a pre-existent symmetry) can it be properly incorporated, enabling us to speak meaningfully of a self-transforming symmetry process.

This leads to a view of cosmogony quite different from the conventional one. Guided by the new intuition, our conception of what constitutes primordiality is altered. With symmetry losing its primordial status, the basic problem of cosmogony changes from that of explaining symmetry breaking to that of understanding how symmetry is created in the first place. The *making* of spatial symmetry, the generation of dimensionality, becomes cosmogony's foremost concern.

In a speculative paper on dimensional generation, Musès (1975) attempted to carry intuition beyond the classical concept of integer dimension. To that end, he introduced the notion of the "fractional dimension,"[1] illustrated by the idea of a "partially generated line" (p. 17). Consider the relationship between the point and the line. While classical thinking demands that we assume the line to be infinitely rich in points, Musès would have us imagine this condition of infinite density being reached only after the line has been *generated* from the point. We are to intuit intermediary stages of production wherein the point density of the line is *less than* infinite, though greater than one. In proposing the idea of a fractionally generated line (not to be confused with the fully generated line segment, which *is* infinitely dense, however short), Musès noted "that the concept of point density bridges the otherwise separated and un-unified concepts of point and line" (Musès 1975, p. 17).

How in particular is dimension $n + 1$ produced from n? By what process is point density increased to create a line? According to Musès, the simplest and most parsimonious solution is an evolving sine-wave function. The wave function, $y = h \sin (2 \pi t/\lambda)$, is given, where t is a space interval, h is wave amplitude, and λ is wavelength. The function is diagrammatically expressed in Figure 6.1 (which simplifies Musès's [1975, p. 18] original diagram). Musès commented that the "only 'real' or manifest portions of the dimension are the points on the wave axis at half wave-length or $\lambda/2$" (p. 18). Now we are to picture the wave form changing so that "D approaches unity, i.e. ...a part-line becomes a line. As this happens, the frequency of the wave markedly increases and its amplitude decreases till at $D = 1$ the wave-length is zero (i.e. infinite frequency) as well as the amplitude" (p. 18). In other words, since point density is directly proportional to wave frequency, an increase in the latter "fills in" the line.

However, while the general notion of dimensional development constitutes a challenge to the classical intuition of the spatial continuum, Musès's particular formulation of it actually presupposes continuity. Musès's evolving sine wave is a *continuous transformation*. To posit such change is, at the same time, to posit the more fundamental invariance of

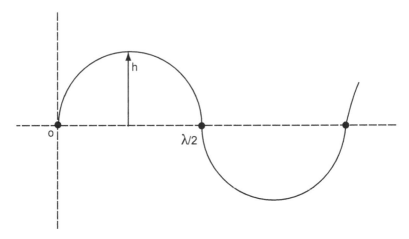

Figure 6.1. Dimensional wave (adapted from Muses 1975, p. 18)

the analytic context in which it occurs. The transformation of the wave form by means of which the line is "generated" assumes the topological invariance of the two-dimensional plane of analysis. Density variations in the line are not dimensionally intrinsic but merely result from the topological treatment the line is given in invariant two-space. That is, the "density varying" line is treated as an *objectified* dimension embedded in an invariant epistemological space (see Chapter 4). Implicitly operative here is the old formula of object-in-space-before-subject, with the "part-line" playing the role of object, not of space per se. Thus, since the "part-line" is but an objectified dimension, its less than infinite density has no bearing on dimensional density as such.

The limitation of Musès's approach may be understood from another perspective, by considering the mistake of assuming that "a partially generating line should contain...some finite number of points" (Musès 1975, p. 17). Reflection discloses that the problem of generating dimension n is not a quantitative one of increasing the number of already well-formed boundary elements in the manifold of D_n to infinity; rather, it involves the problem of creating the *quality* of boundedness in the first place. If a boundary exists that permits one sharply to distinguish any point from another, in effect, we already possess the full-blown, infinitely dense line; what is required for the line is this dimensional quality of two-pointedness. Distinctions among numbers of points beyond two therefore are irrelevant to the question of dimensional generation. This suggests that, instead of being composed of a finite number of already well-defined points, a fractional line is a more "primordial" entity, one whose boundary element is not yet completely formed, has not yet fully been sharpened into focus. (This, of course, is the problem with which Menger was grappling when he abandoned his "topology of lumps" to contemplate "something still more vague.")

In sum, while Musès introduced the notion of "dimensional generation," he tacitly was accepting the classical intuition of invariant integer dimensionality. In Chapter 3 we found that the classical idea of space goes hand in hand with a mechanistic conception of the objects that occupy said space. On such a view, objects are seen to be assembled from the aggregation of externally related, ready-made elements or building blocks (like the assembly of a tinker toy). It is not surprising then that, in Musès's objectified linear dimension, the line should be conceived as being built up from the accumulation of already well-formed, externally related point elements. And in employing dimensions with already formed identity elements, Musès failed to meet a primary requirement for an adequate concept of dimensional generation: one must account for the generation of the identity elements themselves.

6.3 The Three Basic Stages of Dimensional Generation

It is phenomenological intuition that is needed to gain a genuine understanding of dimensional generation. In particular, we have found that Merleau-Ponty's depth dimension provides us with an insight into the embodied dialectical interplay of object, space, and subject lying behind the Cartesian facade of a changeless space wherein objects are cast before detached subjects. In the world of depth, dimensional identity elements are not merely structureless, externally related units that form a static spatial context for the mechanistic transformation of externally related objects; rather, the unit of lifeworld dimensionality is a "spatio-sub-objective" being (what Merleau-Ponty called "an 'element' of Being"; 1968, p. 139) that transforms itself organically. We can see from this that the phenomenological approach lends itself to the idea of dimensional generation in a way that classical thinking does not. The depth dimension does not just *contain* earthly matter but is itself of the earth.[2] Like earthly matter, depth is in process. And we are about to see that, like living matter, it passes through stages of biogenesis. Depth thus possesses the character of a living organism — though not a *finite particular* organism, to be sure. Rather, depth is a generic organicity, a whole dimension of life; neither object nor subject alone, it is a sub-objective lifeworld. Nonetheless, that world develops.

I opened Chapter 1 by relating the quest for unity in science (for symmetry, invariance, or continuity) to *human* development. Here we saw that development is guided by the tendency toward individuation: the drive to be undivided or 'in-dividual,' to possess a unitary core, a coherent, stable, and well-bounded center of identity. We found, however, that the individual does not develop in isolation, but in dialectical relationship to others, and to space itself. The individuation process thus entails the three interrelated aspects of subject, object, and space. We know too that, in the case of the particular human being, this process commences around six months of age with the mirror stage. As Grosz puts it, the "mirror image provides an anticipatory ideal of unity to which the ego will always aspire" (1994, p. 42) — even though this "model of bodily integrity" is something "the subject's experience can never confirm" (p. 43). However, while "the stability of the unified body image...is always precarious" (p. 43), it is just this imagined unity of the ego or subject that anchors particular human identity and permits the process of human development to go forward.

If the mirror stage is the period during which "the division between subject and object (even the subject's capacity to take itself as an object) becomes possible for the first time" (Grosz 1994, p. 32), it indeed marks the beginning of individuation. But it is not the beginning of *Individuation*. By capitalizing the latter term, I mean to indicate that here we are not just dealing with the development of a particular ego or subjectivity but of a cosmic organism encompassing the full dialectical intimacy of subject, object, and space itself. In turning to the question of Individuation, our concern is with *dimensional* development, with the generation of the lifeworld per se.

To understand the developmental process in depth-dimensional terms, it will not suffice to go back to the young child's first experiences with objects in space. We must go still further backward into infancy. At the very outset of development, there is no "infantile observer," no immature Cartesian subject who may be epistemically fused with its objective environment in that it can *perceive* no differences between itself and others, but is nonetheless *ontologically separated* from that environment. Ontologically, what we are dealing with initially is the opening phase in the development of the lifeworld as such, a stage in which a subject has simply not yet been separated from an object, that wherein the projection of object-in-space-before-subject has not yet occurred. In the primal situation then, it is not just a matter of the subject being unable to *know* the object as distinct from itself; rather, subject and object are *ontologically* undifferentiated here; they do not yet constitute well-formed modalities of the depth dimension.

We may say that in the embryonic stage of dimensional generation, the dimensional organism is unoriented, its subject-object relationship lacking direction. Then, in the ensuing stage, subject and object begin to be differentiated by conferring a direction upon their interaction. With the projection of object-in-space-before-subject, awareness is now oriented to function in a single direction: *from* the subjective ground of reflection *to* the objects that are cast before the reflecting subject (see Chapter 1). Obscured by the subject-to-object movement of reflection in idealized Cartesian space is the *pre*reflective action of the "natal space" (Merleau-Ponty 1964, p. 176), the lifeworld dimension that — in the course of its development — *gives birth* to object-in-space-before-subject. In its autopoietic action, it is the dimensional organism that first opens up the field of objectification, and, in the same stroke, first gives a subject who detachedly reflects upon the objects in this field. The initially unoriented lifeworld thus affords itself direction by transforming itself into intentional subjectivity, subjectivity-in-reflection-upon-its-field-of-objectification. In the process of clarifying itself in this manner, the lifeworld *conceals* itself. It facilitates its own development by moving away from itself. In this second stage of Individuation, the dimensional organism gives itself reflective clarity by means of a prereflective process that is itself excluded from the giving.

It is important to recognize that the lifeworld's self-clarification is not deliberate or even conscious in this stage of its development. If it seems strange that the dimensional organism could both clarify itself and conceal itself without being consciously aware of doing either, then we must be clearer about the difference between prereflective and reflective ways of acting. It is the latter that entails conscious acts of deliberation carried out by a subject. Intentional subjectivity is what the lifeworld *gives*, but the giver itself is not a conscious subject. Perhaps we can characterize prereflective lifeworld action as "instinctive": "impelled by an inner or animating agency; hence, imbued; filled; charged; as, a poem *instinct* with passion."[3] The *modus operandi* here is "natural, voluntary, spontaneous... impulsive."[4] "Wild Being" (Merleau-Ponty 1968, p. 211) is no coolly deliberative *cogito*.

What of the *third* stage of Individuation? If, in the second stage, the lifeworld gives itself reflective clarity in such a way that its prereflective giving is eclipsed, what the third stage basically calls for is a *conscious acknowledgment* of this lifeworld action that *integrates* the reflective and prereflective. As we saw in previous chapters, uncovering the hidden dimension of depth requires a "switching of gears," a phenomenological reversal of the prevailing orientation of thinking. Instead of moving out of and away from the lifeworld as happens in stage two, we move backward into it, thereby drawing back in the clarifying reflective light that prereflectively had been emitted, withdrawing its projection. Putting it a little differently, to enter the third stage of dimensional development, we must function *proprioceptively*.

Let us contrast this mode of action with the better known symbolic operations of perception and conception prevalent in conventional science. Etymologically, to perceive is to "take hold of" or "take through" (from the Latin, *per*, through, and *capere*, to take), and to conceive is to "gather or take in." These activities are carried out in the "forward gear," the orientation dominant in stage two. The term *proprioceive* is from the Latin, *proprius*, meaning "one's own." Literally, then, proprioception means "taking one's own," which can be read as a taking of self or "self-taking." It is true that the term's conventional meaning derives from physiology, where it signifies an organism's sensitivity to activity in its own muscles, joints, and tendons. But Bohm (1994) spoke of the need for "*proprioceptive thought*" (p. 229), which he viewed as a meditative act wherein "consciousness...[becomes] aware of its own implicate activity, in which its content originates" (p. 232). Years earlier, the social psychiatrist Trigant Burrow spoke similarly of the need for human beings to gain a proprioceptive awareness of the organismic basis of their divisive symbolic activity (see Galt 1995). What I propose here is that a proprioceptive move is required if the dimensional organism is to gain the awareness of itself that is necessary for entering the climactic stage of its Individuation. Whereas the forward orientation of stage two maintains the trichotomy of object, space, and subject, the proprioceptive unearthing of the lifeworld carried out in stage three would disclose through its "backward" action the intimate transpermeation of these three terms.

The three basic stages of Individuation have now been identified. They may be summarized as follows: (1) non-orientation, the embryonic stage of organismic dimensional generation wherein subject and object are largely undifferentiated; (2) the stage of dimensional autopoiesis in which subject and object are differentiated via a "forward"-oriented action exclusively directed from the former to the latter; in thus clarifying itself, the lifeworld moves out of and away from itself; (3) the climactic "reversal of gears" in which the lifeworld returns to itself via the proprioceptive circulation wherein subject and object flow unbrokenly into one another. Of course, the depth dimension previously has been fleshed out by means of topological phenomenology. Since our opening description of dimensional generation does not take this into account, the task presently at hand is to provide a topodimensional rendition of the stages. We begin with some preliminary considerations.

According to classical intuition, the spatial continuum is bounded by an infinity of juxtaposed planes, elements that are themselves unextended in three dimensions and thus possess no internal structure of their own. As noted in Chapter 1, the elements of classical space are not internally substantial, concretely bounded entities but constitute a condition of abstract boundedness as such. It is within the infinite boundedness of space that *particular* boundaries are formed, boundaries that enclose what is concrete and substantial. The concreteness of what appears within planar boundaries is the particularity of the object. To reiterate, the object is that which is bounded, and space, via its infinite reservoir of planar bounding elements, provides the context that enables the object thus to be differentiated.

Bearing in mind our present concern with topodimensional generation — or shall we say, *topogeny*, let us sharpen our focus on the nature of dimensional boundedness. Do the dialectically opposed surfaces that bound an object in space derive from and reduce to the inert juxtapositional relation of structureless planes bounding space per se, as is implicit in classical thinking? We have seen that the answer is no. With the help of Merleau-Ponty, we have come to understand that, on the contrary, the dialectic goes "all the way down." As a consequence, the boundedness of space must itself undergo dialectical development.

It will prove helpful to state this proposition in terms of the issue of symmetry. Recall the distinction made in Chapter 2 between object symmetry and the symmetry of (analytical) space itself. Mathematically, an object is symmetric to the extent that it remains the same when some specific transformation is introduced (e.g., a sphere possesses perfect rotational symmetry because its appearance does not change however it may be turned upon its axis). We may say that, in this kind of symmetry, invariance is relative in nature. For it is not that the object is not transformed at all (in topology, we can even change coffee cups into doughnuts), but that concrete changes in the object are offset by a substrate of changelessness (the coffee cup and doughnut are entirely equivalent from the abstract standpoint of their continuous inter-convertibility). However, while the symmetry of an object in space is in this sense relative, it is dependent upon symmetry of a stronger sort: that of space per se. Whatever structural variations may arise in the objects that are embedded in space, the spatial continuum itself — assumed to be composed of infinitely many structureless bounding elements all inertly packed together — remains *absolutely* invariant. So, on the classical view, space is complete unto itself and totally impervious to change. Exactly why is the absolute symmetry of the spatial container assumed to be the precondition for all relative symmetries in the objects contained? It is because the bounding or containment of change that classical object symmetries entail requires the forming of particular boundaries from the infinite boundedness that is space.

In stark contrast to the closed juxtapositional stasis of the idealized symmetric continuum is the openness of the actual lifeworld. It is in participating in the lifeworld that we encounter the dialectical phenomena of "orientation" and "polarity" of which Merleau-Ponty spoke (1964, p. 173); that is, lifeworld experience is not blandly homogeneous but distinctly directed, oriented, inherently susceptible to the development of polar opposition. Here is where "there are no events without someone to whom they happen and whose finite perspective is the basis of their individuality" (Merleau-Ponty 1945/1962, p. 411). Of course, on the classical reading, the perspectival finitude of the lifeworld, its incompleteness or asymmetry, is merely a surface appearance concealing an underlying condition of symmetry. When Merleau-Ponty *questioned* the classical idea that lifeworld asymmetries are just "derived phenomena"; when he challenged the notion that "space remains absolutely in itself, everywhere equal to itself, homogeneous" (1964, p. 173); when, for classical space, he offered the more primordial dimension of *depth* — he was exploring the possibility that nonsymmetric dialecticity indeed goes all the way down into the roots of space.

We saw in the last chapter that classical idealization hinges on denying phenomenological fact. For example, it is a fact of perspectival perception in three-dimensional space that we cannot see all six surfaces of a cube at once. We can simultaneously view all four lines bounding a two-dimensional square. We can perceive at once both of the end-points that bound the one-dimensional line segment. Yet when it comes to the six surfaces that bound the cube, we are limited to simultaneously viewing a maximum of *three* of them (along with their angular relationships; see Fig. 5.5). But the classical idealization glosses over this asymmetry by offering us abstract reassurance that, despite the *appearance* of the cube's incompleteness, the whole cube is "really there" at the moment we are viewing it; that its opposing perspectives are at bottom *juxta*posed in simple simultaneity; and that the space in which the cube is embedded remains a seamless continuum whose absolute symmetry is upheld. To Merleau-Ponty, however, phenomenological reality must take precedence over abstract ideals.

But does phenomenological reality dictate that space is *merely* asymmetric, that it is simply *dis*continuous, that its bounding elements are simply open, incomplete? What I am proposing is that the lifeworld undergoes *development*, that it transforms itself through several stages of Individuation. Given that this organic world is dimensional, the stages of Individuation must entail dimensional self-transformation. It will therefore not suffice to replace the classical idealization of the simply closed continuum, of simply symmetric boundedness, with the merely asymmetric notion of the incomplete or open boundary. We require instead an understanding of the full course of dimensional development whereby the boundedness of space — including its classical idealization — is dialectically self-generated. Transposing these terms, we can speak of the generation of *self-boundedness* or self-containment. This way of describing dimensional generation highlights the idea that the lifeworld is an organism whose Individuation entails the realization of its self-containing character. I submit then that lifeworld dimensionality requires us to think in terms of a process of transformation by which the dialectical development of spatial self-boundedness occurs.

With this in mind, we shall now inquire into the topological characteristics of the general stages of Individuation set forth above. For the remainder of the chapter, the account will focus on the lifeworld whose manifestation is most familiar to us: the world of three-dimensional space, which derives from Kleinian quantum-vortical action (ε_{D3}). In subsequent chapters, we will deal with the topodimensional development of the hidden, lower-dimensional lifeworlds (Chapter 7), and with the forces of nature to which the several lifeworlds are related (Chapters 8–10).

6.4 Kleinian Topogeny

Since the phenomenologically-interpreted Klein bottle embodies the three-dimensional lifeworld, the lifeworld's organic development can be understood in terms of the Klein bottle's development. We have found that the construction of a Klein bottle can be approximated by gluing together oppositely oriented, asymmetric Moebius strips. Of course, operations with paper models lend themselves to the old formula. The assumption is that we are working with topological objects situated in space, and that we Cartesian operators are aloof from the objects we manipulate. But the *necessary hole* that is produced in attempting to form the Klein bottle as an ordinary object in space (Moebius strips cannot be glued together without a tear) leads us to the conclusion that the proper generation of the Klein bottle does not actually involve mechanical operations upon finite particular objects embedded in a static space. In its capacity as a dimensional organism, the Klein bottle's development in fact entails the prereflective evolution of space itself, which is inseparably linked to the dialectical evolution of generic subjectivity and objectivity (the manifestation of *particular* subjects and the objects upon which they reflect occurs in the second stage of Individuation, and arises from the prereflective cosmic matrix). We need to realize then, that our mechanical model of the formation of the Klein bottle from Moebius enantiomorphs signifies a process that itself is not mechanical but generically embodied.

We already know that Kleinian self-circulation involves a dialectic of objectivity and subjectivity, symmetry and asymmetry, wholeness and "holeness." Now, in focusing on the *diachronic* aspect of this, we will trace the stages of Individuation through which the spatio-sub-objective Kleinian organism develops via fusion of lower-dimensional enantiomorphs. Despite the fact that the enantiomorphs appear in the mechanical model to be merely positive, simply continuous entities, particular objects in space, we will regard them as generically organic since it is they that facilitate the generic Kleinian organism's maturation. Near the end of the previous chapter, I suggested that one enantiomorph, denoted as $\varepsilon_{D2}/\varepsilon_{D3}$ in Table 5.2, operates as a "fractional" (incipient) form of the Kleinian wave, an "undertone" of it, that is engaged in dimensional self-generation; the reciprocal enantiomorph, $\varepsilon_{D3}/\varepsilon_{D2}$, functions in the capacity of a Moebius carrier wave, an "overtone" of the Moebius that guides Kleinian autopoiesis. When the merger of enantiomorphs is completed, they are annihilated as free-standing entities. The Moebius "overtone," having fulfilled its role as carrier wave, is absorbed into the three-dimensional Kleinian vortex; at the same time, the nascent Kleinian "undertone" is brought to term as ε_{D3}, the fully three-dimensional Kleinian vortex itself. With this internalization of free-standing enantiomorphic waves to the integral Kleinian wave, they now function as enantiomorphically related elements of the Kleinian vortex (the distillation of bounding elements from enantiomorphic fusions will be considered in the next chapter).

Now, as products of the second stage of Individuation, mechanical models presuppose external relations among the entities being modeled. Accordingly, in the model portraying the formation of the Klein bottle from Moebius enantiomorphs, the process of enantiomorphic fusion begins with already well-differentiated Moebius strips that are sharply set apart from each other. In depth-dimensional terms, however, the generation of the Kleinian structure begins at an earlier stage. The opening phase of dimensional generation consists of a primordial situation in which enantiomorphs that are to be fused are as yet neither differentiated from one another nor integrated (see Chapter 4 for a discussion of the dialectical relationship between differentiation and integration). In this embryonic stage of Kleinian autopoiesis, no boundary as yet exists that would define enantiomorphs in opposition to each other. It is as if enantiomorphs that will eventually come together are at first "infinitely distant" from one another insofar as they have not yet even been brought into the same frame of reference where their differences can be defined. This incipient condition of boundlessness denotes a Kleinian space of zero density; since there is as yet no bounded continuum, we may associate the stage with pure discontinuity, as though only the hole in the Kleinian vortex exists and not its positive elements. We may conclude then that dimensional development commences from a nascent state of "pure holeness," one in which the opposition of hole and whole — subject and object, does not exist. Should we say that the Kleinian boundary is utterly open in this nascent phase, since the process of realizing spatial self-boundedness via fusion of enantiomorphs is still to begin? It seems what we must actually say is that, while the dimensional vessel indeed has no closure because enantiomorphs have not yet begun to merge, neither does it possess any openness inasmuch as enantiomorphs have not yet begun to be separated either. A logic of "neither/nor" thus appears appropriate for the incipient phase of dimensional generation. (In Chapter 5 we discovered that the Kleinian dimensional wave is enantiomorphically carried not only by the two-dimensional Moebius wave but by two other dimensional vortices as well. The co-evolution of all dimensional waves will be explored in the next chapter.)

We have seen that the lack of differentiation of subject and object prevailing in the first stage of Individuation is coupled with a lack of direction or orientation, and that, in entering stage two, subject and object become differentiated in such a way that all action is directed exclusively "forward," from the former to the latter. This can now be understood topodimensionally, in terms of the fusion of enantiomorphs involved in the genesis of the Kleinian dimension. In the second stage, enantiomorphs are brought together to the extent of becoming differentiated within a common frame of reference wherein action is unidirectionally oriented *from* the generic subject enantiomorph *to* its object counterpart. Stage two is characterized by covert dimensional asymmetry. That is to say, while dimensional boundedness is actually incomplete in this phase, it is *projected* as complete by positing the ideal of the perfectly symmetric, infinitely bounded continuum — along with the perfectly self-symmetric, autonomous subject that exerts its influence upon the objects in this space without itself being influenced. The underlying fact, however, is that another stage will be required for the Kleinian lifeworld to realize full self-boundedness, a stage in which the merger of enantiomorphs is completed in earnest synsymmetrically via the switching of gears from "forward" to "backward" (see below). Here the dimensional organism that had concealed itself in stage two now surpasses the egoic idealizations of this stage to gain the proprioceptive awareness of itself that is necessary for achieving its Individuation.

Is it really accurate to say that, in the intermediary stage, enantiomorphic fusion is only *partial*? While this is true in a general sense, stage two fusion is not *simply* incomplete but is distinguished by the fact that its incompleteness is masked and the appearance of completeness is projected. Therefore, rather than saying that stage two in the generation of dimensional self-boundedness entails a *partial* fusion of enantiomorphs, a partial sealing of the dimensional vessel, it appears more appropriate to describe this as a *quasi*-fusion or closure: the vessel *seems* to be sealed but it actually is not. In fact, there is a topological model of dialectical process that displays this feature of quasi-closure in a graphic fashion: the Moebius strip.

Movement about the one-sided Moebius surface involves the traversal of two cycles. This can be illustrated by returning to our comparison of the Moebius and cylindrical ring (Fig. 5.1). Consider a circulation about the cylindrical ring (Fig. 5.1(a)). Positioned on the inner surface of this simply two-sided structure, we move 360° around to complete one revolution, returning to our point of departure. Naturally, our passage around the inside of the ring never takes us to the outside. Throughout the journey, we remain on the side on which we began.

It is not like this with movement around the one-sided Moebius surface (Fig. 5.1(b)). While a 360° revolution does seem to bring us back to our point of origin, at the same time, it is *not* our point of origin, since we are now on the *other side* of the surface. Notice the way ordinary cylindrical revolution maintains the simple dichotomy between point of origin and displacement from that point: by 180° of movement on the cylindrical ring, we are furthest removed from where we began, and by 360°, we are back where we started, the displacement being simply and completely reversed.

In contrast, with a 360° turn around the Moebius, there is a circling back to the point of origin that at once is the point *most remote* from said origin, since — when sidedness is taken into account — it is the point on the other side of the surface that is diametrically opposed to the actual starting point. What happens if we continue our movement beyond this 360° point of quasi-return? After an additional 360°, we find that now we have truly returned, since we have come back to our genuine point of departure, that on the side of the surface from which our movement had commenced.

The double Moebius cycle thus essentially involves a quasi-return to origin, followed by an additional circuit that repeats the process, now completing it in earnest via a second return. In this way, Moebius action models the dialectical process of dimensional generation in which there is first a quasi-closure, a quasi-sealing of the dimensional vessel, and then an authentic closure or act of containment. Clearly related to this, we can also see how the initial closure of the Moebius circuit symbolizes the quasi-fusion of enantiomorphs indicative of the intermediary stage of the dialectic, whereas the second closing of the Moebius circuit corresponds to the realization of full-fledged enantiomorphic fusion associated with the final stage of dimensional development. The sides of any topological surface are related to each other as mirror opposites. In the case of the simply two-sided cylindrical ring, a full 360° turn naturally leaves us on the same side on which we started; opposites thus go unreconciled. In contrast, the Moebius dialectic brings mirror-opposed sides together. In the initial Moebius circulation, this happens only to the extent of completing the passage from one side to the other. At this point, both sides have not been traversed in their entirety; in fact, only *half* of the one-sided surface has been covered. It is the *complete* encompassment of enantiomorphically related sides that is required to fulfill Moebius one-sidedness, and, for this, there must be the additional Moebius circuit.

Note a correlated feature of the Individuation process that is modeled by Moebius action: the stage-three "switching of gears" from "forward" to "backward"; from the projective idealizations of classical intuition to the proprioceptive reality of phenomenological intuition. To demonstrate the property of gear reversal, we incorporate a test body into the Moebius model: an asymmetric profile.

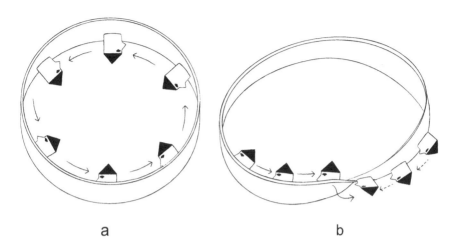

Figure 6.2. Revolution of asymmetric figure on cylindrical ring (a)
and on Moebius strip (b)

Figure 6.2(a) shows a left-facing profile revolving in a counterclockwise direction around a cylindrical ring. It is clear that action on the simple ring will continue indefinitely in this manner, with the orientation of the profile never changing (though the half-face is turned upside down). In Figure 6.2(b), we see the profile moving counterclockwise about the Moebius strip. Entering the twist, the left-facing form is changed into a right-facing form, the transformation being completed after 360° have been traversed. The transformation of left into right thus coincides with the occurrence of one full cycle of Moebius action. This change in orientation can be seen to reflect a change in clock sense, for, what is counterclockwise to a left-facing profile will be clockwise to one that faces right. Thus we may say that Moebius action involves a transformation of clock sense, a reversal of the counterclockwise or backward orientation that results in a forward or clockwise inclination. It is obvious that cycle one is not simply clockwise from beginning to end, this being followed by a shift in gears that gives a second, uniformly counterclockwise cycle. Rather, the clockwise orientation of cycle one is realized only at the end of the cycle as the culmination of an ongoing transformation from an initial counterclockwise orientation. Then, in entering cycle two, the direction of the gear shift has itself shifted so that the momentum is now from clockwise to counterclockwise, forward to backward. The fact that change in clock sense is *inherent* to Moebius action means that no *separate* switching of gears is required to bring it about. The gears are shifting throughout each cycle, with the quasi-return to origin marking the completion of the shift in one direction, and the readiness to begin shifting in the other. In this way, cycle one reaches its climax in the forward orientation symbolic of projective idealization, and, with commencement of cycle two, we begin the retrograde or backward movement that signifies proprioception.

I hasten to emphasize, however, that our observations of Moebius transformation provide us with no more than an analogy of the three-dimensional transformations we actually seek to understand. For we know that, in the three-dimensional context, the Moebius strip is naught but an object-in-space. Though the Moebius can *symbolize* lifeworld transformation here, it lacks the special, self-intersective property required for playing out the dialectic of opposing sides of surfaces. Therefore, to embody Individuation in the three-dimensional milieu, it is the one-sidedness of the *Klein bottle* to which we must turn.

Like the Moebius that is involved in its formation, the Klein bottle should consist of a double cycle. That is because, as with the Moebius, two cycles should be required to fully express the dialectical property of one-sidedness, whereas the simply two-sided, non-dialectical counterpart of the Klein bottle, viz. the torus (Fig. 4.3, upper row), entails but a single cycle of 360°. With this in mind, imagine a circulation that commences from a point on the inside of the Klein bottle and proceeds for 360° about the length of the structure (see Fig. 4.1). It seems that this revolution should produce the quasi-return that brings us back to our point of departure, but on the outside (similar to what happens with 360° of Moebius action). Continuing movement along the same trajectory for an additional 360°, we should then regain our true point of departure on the inside of the bottle. However, unlike what happens with Moebius revolution in three-dimensional space, the continuity of side-to-side Kleinian circulation is broken by the occurrence of the self-intersection: to pass from inside to outside and back again, the surface of the bottle must be breached. It is this interruption of mechanical action that invites a *phenomenological* understanding of the double cyclicity of the one-sided Klein bottle. Nevertheless, we have seen that conventional topology has its way of repressing the challenge to continuity posed by Kleinian one-sidedness. It may be helpful to clarify a little further the standard approach.

Conventional topology treats the structural features of a surface (its curvature, for example) as global properties resulting from the way this mathematical object is embedded in space. Whatever may be the global structure of such an object, the spatial continuum itself is assumed to consist of local bounding elements that are structurelessly juxtaposed. To conventional thinking, the topological merging of the sides of a surface is a perfectly permissible operation, as long as it can be understood as a merely global effect, a transformation in which local (point-to-point) continuity is not compromised. By the same token, a surface's "non-orientability" is acceptable as a strictly global feature. Although the fusing of sides brings a "switching of gears," a change in left-right or clock orientation involving asymmetric opposition, this poses no problem for the standard approach if the operation can be seen as having no effect on the self-symmetric bounding elements of the juxtapositional continuum.

Of course, in the case of the Klein bottle, it is not actually true that the topological transformation can properly be regarded as a global event that conserves the local continuity of three-dimensional space. Since Kleinian transformation requires negotiating the *hole* it produces in itself, we in fact cannot trace a simply continuous path on the Klein bottle from one side to the other. In the classical approach, this crucial discontinuity is avoided by the questionable act of abstraction we already have examined: the hole in three-dimensional space that prevents continuous Kleinian transformation is circumvented by gratuitously invoking a four-dimensional continuum in which the Klein bottle is purportedly embedded. It is when we resist the inclination to proceed in the classical fashion that we are led to the phenomenological realization that the Kleinian phenomena of "orientation" and "polarity" — the dialectical interplay of left and right, backward and forward, inside and outside — do not merely involve the global dynamics of an object embedded in static space but mirror the prereflective core dynamics of space itself, of the depth-dimensional lifeworld. Moreover, these transformations are consistent with the *development* of the lifeworld described above.

The phenomenological interpretation of the cycles of Kleinian action departs significantly from the classical picture of point-to-point movement through the spatial continuum. In the mechanical approach, rotation about the Klein bottle commences from a point on one of its sides, the implication being that, locally, the Klein bottle does possess two sides. The contrasting phenomenological view tells us that, at the outset, we do not have the condition of local two-sidedness so compelling to classical intuition but rather, the primordial state of affairs in which enantiomorphic opposition between Kleinian sides does not yet exist. Moreover, phenomenologically, "opposing sides" of the Klein bottle do not involve the sides of a topological object-in-space. It is objectivity and subjectivity themselves that are neither differentiated nor integrated in the opening stage of Individuation. What happens with entry into stage two, the stage expressed by commencement of the first cycle of Kleinian dimensional action? Contra the mechanical model, we do not merely have a continuous movement from one locally well-differentiated side of the topological surface to the other. Instead we have the *differentiation and integration* of "sides," that is, of subject-object enantiomorphs. The process unfolds in such a way that a distinct orientation arises, a unidirectional forward movement *from* subject *to* object, as we have seen. So it is not the mere *switching* of orientation that occurs in cycle one as suggested by the mechanical model, but the *establishment* of a definite orientation from an initial condition in which orientation is lacking. (Bear in mind the discrepancy between cycle number and stage number: cycles one and two of topodimensional action correspond respectively to stages two and three of Individuation, with the first stage entailing no directed cyclical action.)

Above I described the second stage of Individuation as involving a quasi-closure or integration of the dimensional vessel. What is presently clear is that this quasi-integration of enantiomorphs is just as much a quasi-*differentiation*. For, in moving from stage one to stage two, we do not just go from openness to closure, but from an initial situation in which there is *neither* openness *nor* closure, to one in which there is *both*. These features can be studied in more detail via our topological models.

Consider once more the torus (Fig. 4.3, upper row, and Fig. 5.6(c)). This higher-order counterpart of the cylindrical ring possesses two distinct planes of revolution, longitudinal (L) and transverse (T) (Fig. 6.3). The planes of action of the torus are orthogonal, simply independent of one another. Just as the earth revolves around the sun, and, at the same time,

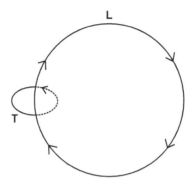

Figure 6.3. Orthogonal circulations of the torus: longitudinal (L) and transverse (T)

rotates upon its own axis, we can imagine perpendicular toroidal circulations to be occurring simultaneously. Yet any combination of these actions
is strictly linear: their respective contributions add together without an
internal change in either.

Let us take the longitudinal action plane of the torus as our plane of
reference and regard it as a topological object with opposing sides. A 360°
orbit about one side of this plane would be restricted to that side and would
describe a circle of simple self-return. For its part, the 360° transverse
circulation, the revolution occurring at right angles to the reference plane,
would constitute an orbit of "simple departure" insofar as it would bring
us to the diametrically opposite side of the reference plane. If we regard
the longitudinal "circle of return" enacted on the given side of the reference plane as signifying closure or boundedness, self-identity or integrity
(integration), the transverse "circle of departure" would then express
opening, the transgression of boundaries, difference or differentiation. The
strictly dichotomous relation of toroidal action components thus symbolically expresses the dualism of difference and identity. The logic of *either/
or* is embodied here, for even though difference and identity operations
may be thought of as occurring "at the same time," any *single* operation
must involve *either* difference *or* identity, given the external relation between these opposites. Thus, in the single operation, there will either be
the self-return taking place on the same side of the reference plane, or the
transverse displacement to the other side, but never both. Moreover, this
dualism of identity and difference is accompanied by a dualism of com-

pleteness and incompleteness. That is, whether the action is longitudinal or transverse, it will be either simply complete or simply incomplete. When 360° are spanned, the action is complete; for anything less, it is incomplete.

Returning to the Moebius model, we find a contrasting action pattern, one in which longitudinal and transverse components are thoroughly blended. Yet can we not identify two distinct circulatory components of Moebius action? Consider again the movement of the asymmetric profile upon the Moebius strip (Fig. 6.2(b)). The profile orbits the length of the Moebius and also rotates upon its own axis in the direction perpendicular to the plane of longitudinal revolution. The longitudinal component of Moebius circulation is what brings the profile back to its starting point after 360°; this is the aspect of return, that which signifies closure, integration, boundedness. The transverse Moebius component, which entails only 180° of action, carries the profile to the *other side* of the surface; in so doing, it constitutes the aspect of departure that symbolizes opening, differentiation, and the surmounting of boundaries. However, unlike the action components of the torus, the circulations of the Moebius are interwoven in such a way that the return is at once a departure, and the departure a return! It is because the first cycle of Moebius action combines return and departure (identity and difference) in an integral way that a second cycle is required to fulfill these operations.

Under the continuing influence of classical logic, we might persist in asking whether there is or is not a return to origin after one cycle of Moebius action. Yes or no, we may inquire. The answer, however, is yes *and* no. The circle of identity is indeed closed with 360° of Moebius action. But because this circulation upon the same side of the longitudinal plane is inextricably coupled with the transverse displacement to the other side, the return to origin must be enacted again. It is for this reason that the cycle-one return is a *quasi*-return. In cycle two, the return movement is completed in earnest by reversing the departure that took place in cycle one: by a second transverse action, the profile now reverts to the side of the Moebius from which it began.

The first Moebius cycle constitutes a quasi-departure as well as a quasi-return. With 180° of rotation, we do have more than a mere *semi*-departure bringing us halfway round to the other side of the longitudinal action plane. Rather, the 180° transverse component of Moebius action brings us *all* the way around to the other side. And yet, since this transverse circulation is integrally combined with the longitudinal circulation that keeps the profile on the same side of the plane, further action is required in cycle two for departure to fully and truly be realized. Coupled with the transverse rotational component of cycle two that reverses the transverse action of cycle one by returning the profile to the side of the surface from which it had embarked, there is a *longitudinal* element that, in effect, offsets the longitudinal action of cycle one that had then limited the act of departure. Whereas the first cycle of longitudinal action had had the effect of maintaining the profile's position on the original side, in the second cycle longitudinal action serves to maintain the profile's position on the side to which it was displaced by cycle one transverse action. In this way, the quasi-departure of cycle one becomes full-fledged departure in cycle two. The paradox, of course, is that the departure *is* the return (and vice versa); the true differentiation of sides is at once their true integration.

Transposed to the Klein bottle for a depth-dimensional reading, the quasi-differentiation and quasi-integration characteristic of cycle one constitute the state of affairs prevailing in the second stage of Individuation, where it is *subject* and *object* that are thus differentiated and integrated. In the quasi-complete fashion, dimensional enantiomorphs fuse, come together in an act of self-bounding that weds the emergent subject to the immanent world of objects; by the same paradoxical stroke, enantiomorphs move apart, so that the subject gains quasi-autonomy, achieves quasi-transcendence. Thus there arises the dialectic of constraint and freedom, containment and unboundedness. The two are not just joined at the surface but compounded in depth, blended so thoroughly that they cannot be teased apart.

Stage two is further distinguished by the fact that its dialectical entwinement and quasi-completeness are not recognized as such. Instead there is the tendency to project simple separation and completeness, as well as simple attachment and incompleteness. The appearance is therefore created of the wholly self-sufficient, fully transcendent subject, on the one hand, and the utterly dependent object, on the other. Adding to this the third term of the classical triad, namely, space, we have once again the classical formula: object-in-space-before-subject. Like the subject, space is projected as complete within itself and unchanging, constituting as it does the infinite boundedness of the self-symmetrical continuum. Only the objects that are enclosed within this spatial container are said to change, but these transitory structures are ultimately regarded as secondary phenomena, not as essential in themselves. In the language of conventional topology, they are merely *global* entities, their dialectical properties having no effect on the local character of space per se. This classical ideal of dichotomy and stasis is what we project upon the dynamic Kleinian blending of subject, object, and space when we assume the Cartesian posture in stage two. The second stage of dimensional development is thus the stage wherein dimensional development itself tends strongly to be denied. Since the very notion of an evolving dimensional organism is exceedingly difficult to grasp here, it will surely be no less difficult to comprehend the idea that the development of said organism is only *quasi*-complete in this second stage (first cycle). To acknowledge that the integration and differentiation of subject and object are incomplete would be to admit the lingering influence of the inchoate state of affairs that held sway in stage *one*, the stage in which *neither* identity *nor* difference were clearly in evidence. It is particularly this remnant of "infantile flux" that classical thinking cannot come to grips with, since it belies the "clean transcendence" of the phallic subject. The "womanish" vestige of primal chaos (known in alchemy as *prima materia*: prime matter or mater) is what the idealizations of the classical mind have always sought to obscure (see Rosen 2004).

Our continuing predilection for idealized projection can explain the relative difficulty we have with one-sided topological structures like the Moebius strip and the Klein bottle. They seem so puzzling and circuitously complex to us, compared with the "straight-forward" properties of their two-sided counterparts. When faced with the quasi-closure of the Moebius strip and Klein bottle, we are strongly inclined to mistake it for the simply self-symmetrical closure of the cylindrical ring and the torus. For the latter pair of topological objects well embody the classical ideals of dichotomy and stasis, of simple completeness. The simplifying idealization imposed on paradoxical one-sidedness can be regarded as a kind of "tunnel vision" or narrowing of viewpoint, as illustrated by our topological models. In the case of the Moebius strip, it is when we restrict our view of this surface to a local cross-section that we eliminate its curious global features and thereby obtain the categorical two-sidedness that defines the cylindrical surface in its entirety. In like manner, narrowing our view of the one-sided Klein bottle yields toroidal two-sidedness. Classical two-sidedness therefore can be considered as but an idealized limit case of the more general Moebial/Kleinian truth we find so difficult to countenance.

But *what about* the stage-two limitative tendency to project an appearance of simple completeness upon a process that in fact is incomplete? Is there a specific *topodimensional* feature that reflects this inclination?

We have seen that passage into the second stage of Individuation brings with it the emergence of a definite orientation from an initial situation in which direction is lacking. The direction of action thus established is the movement of classical consciousness out of and away from the lifeworld, the transition that brings us forward, *from* subject *to* object. It is through this propensity that changelessness and completeness are idealistically projected. Obscured in the process is the underlying fact that this is just a stage of Individuation, one in which the development of the Kleinian dimensional organism is actually only *quasi*-complete. Evidently, the "forward gearing" of dimensional action characteristic of stage two does find specific topological expression. In the mechanical model of the Moebius strip, I used an asymmetric profile to bring out the *vectorial* aspect of the action: the shifting of orientation from left to right, counterclockwise to clockwise, backward to forward, in cycle one — and back again, in cycle two. No such "switching of gears" takes place with simply symmetric cylindrical or toroidal action. Of course, phenomenologically, cycle one does not just entail a linear change in orientation but the *establishment* of a definite orientation to begin with. And this is what we have with the Klein bottle when it is apprehended in the prereflective, depth-dimensional way.

If, in stage two (cycle one), vortical dimensional action is *directed*, and if the momentum of this stage spins the dimensional organism in the "forward" direction, what happens when it passes into stage three, the second cycle of Kleinian action? Observation of mechanical rotation on the Moebius strip indicates that the cycle-one transition from left to right is reversed in cycle two. Exactly what does this mean in depth-dimensional terms?

In the second cycle of Kleinian action, the projection of simple completeness enacted in cycle one is *counter*acted. What arises here is a *retro-*

grade circulation, a movement against the "forward" orientation that had prevailed in cycle one. The "gears" have "shifted" in cycle two so that, instead of moving away from the inchoate origin in concealment of it, the dimensional organism moves backward into it. This movement back into the undifferentiated, unintegrated source is not merely regressive; it does not merely cancel the forward thrust toward differentiation and integration, any more than the cycle-one move toward differentiation and integration cancels out the unindividuated source (though that source does become repressed). For, in neither case are we dealing with the point-to-point, linear kind of movement that takes place in the infinitely divisible continuum. It is when movement is linear that, in arriving at point two, point one is simply left behind. In contrast to this action in classical space, the action *of* space, *dimensional* action, can be said to be "holistic" or "integral," since it entails an indivisible, quantized aspect. Therefore, just as the forward orientation of cycle one does not simply leave behind the original lack of orientation, the backward action of cycle two does not merely nullify the forward action.

Suppose you were handling a textured piece of fabric, one whose fibers were arranged in a certain direction. Is it not by running your fingers *against* the grain of the material that its direction becomes more clearly discernable? Similarly, in the cycle-two backward movement against the grain of cycle one, awareness is gained of the very process of forward-directed projection that transpires in that first cycle. In this way, the projection is retracted, consciously taken back, in the midst of its ongoing occurrence. It is in so acknowledging the forward-oriented activity of cycle one that the dimensional organism engages in *proprioception*. Here, rather than simply going *with* the grain — that is, naively buying into the ideal of perfect integration and differentiation — the organism moves in the retrograde manner that allows it to bear witness to how this idealization is first projected from the inchoate origin that had been thoroughly concealed when movement was geared exclusively forward the "first time around." The projection of difference and identity, of autonomy and integrity, is the gift that the Kleinian organism gives to itself in its quasi-self-clarification. And the "reversal of gears" that takes place in the second Kleinian circulation is the proprioceptive *acknowledgment* of this giving that completes it in earnest.

Let me now summarize topodimensionally the three general stages of Individuation. (Though the concern here is still exclusively with Kleinian dimensional generation and the immediate support it receives from the Moebius, in the new chapter we are about to enter our scope will widen to include the generation of the lower dimensions.) Stage one constitutes an embryonic state of affairs in which $\varepsilon_{D2}/\varepsilon_{D3}$ and $\varepsilon_{D3}/\varepsilon_{D2}$ enantiomorphs, whose relationship prefigures the subject-object relationship, are neither integrated nor differentiated; here space is neither closed nor open, neither bounded nor unbounded, symmetric nor asymmetric. Then, in stage two, enantiomorphs achieve their quasi-fusion, with the three-dimensional Kleinian organism now projecting itself as object-in-space-before-subject. Thus there is the quasi-closure of the continuum, the purported infinite boundedness that constitutes perfect spatial symmetry. And, in sharp separation from the positivity of space and the objects it contains, is the negativity of the subject, of that which is unbounded, asymmetric, utterly open or transcendent. Therefore, whereas the dimensionality of stage one entails neither boundedness nor unboundedness, in stage two we have both, but have them in idealized, irreconcilable opposition to one another. As a consequence, the stage-one logic of neither/nor is supplanted by the logic of either/or.

The either/or of stage two is signified topologically by an idealized structure like the torus, whose two directions of circulation appear simply and completely orthogonal to one another (Fig. 6.3). Above we found that the classical orthogonality of the two-sided torus is but a limit case of the more general "hybrid" truth of the one-sided Klein bottle: it is when we restrict our view of the Klein bottle to a local cross-section (neglecting its full length) that it appears indistinguishable from the torus. Then we may associate the quasi-completeness of the second stage of dimensional generation — the quasi-boundedness and quasi-opening, the quasi-symmetry and quasi-asymmetry, the quasi-Individuation — with a narrowed-down view of Kleinian reality, one that gives it the merely circular appearance of a torus. Note once again that this occlusion of Kleinian dimensionality constitutes a stage in the development of the Kleinian organism itself: to find itself, the dimensional being first must lose itself.

In stage three, the fusion of dimensional enantiomorphs is authentically completed. Through the retrograde movement of this stage, the "dimensional blinders" are removed and the dimensional organism is able to recognize three-dimensional action for what it is: the self-permeative cyclonic circulation of the Kleinian lifeworld. As a result of this disclosure, the dualistic juxtaposition of symmetry and asymmetry at play in stage two of dimensional development is replaced by a synsymmetric blending of these opposites. In the climactic phase of Individuation, the logic of either/or gives way to that of both/and: the depth-dimensional organism is both symmetric and asymmetric — not in Sartre's sense of the mere juxtaposition of contradictories, but in Merleau-Ponty's sense of a dialectical commingling in which opposing terms fulfill each other as they permeate each other (see Chapter 3). In the paradox of synsymmetric depth, we have both *more* symmetry and *more* asymmetry than the classical idealizations of these allow. That is, in its very openness, the dimensional vessel is actually bounded more completely than is the classical continuum, with its alleged "infinite boundedness": the synsymmetric vessel is *hermetically sealed* (to use an old alchemical expression). And in its closure, the Kleinian vessel is more open, for this plenum is suffused in its entirety by void. That is to say, the vessel is a *(w)hole*. In short, the same self-circulation that bounds Kleinian dimensionality, that binds it indissolubly to the lived world, at once sets it free.

Notes

1. The idea of "fractional dimension" should not be confused with Mandelbrot's (1977) concept of "fractal dimension." In the latter, topological dimension is not affected by fractal transformations. Thus, the classical intuition of dimension is not directly challenged.

2. Einstein's general theory of relativity also can be said to associate matter with dimensionality, since this is a theory in which gravitational mass is directly expressed via space-time curvature. However, though Einstein employed Riemannian geometry to describe the global-topological deformation of space-time related to gravitational effects, he clearly did not intend to suggest an intrinsic deformation of the analytic continuum per se. Einstein had assumed that his space, when considered in the small, would flatten out and be homeomorphic to classical Euclidean space. That is why Einstein was so appalled to discover that his own theory led to predictions of *holes* in space.

3. *Webster's New Collegiate Dictionary*, 1975 ed., s.v. "instinctive."

4. *Webster's New Twentieth Century Dictionary*, 2nd ed., s.v. "instinctive."

Chapter 7
Waves Carrying Waves:
The Co-Evolution of Lifeworlds

Examining the overall course of dimensional generation in the preceding chapter, we maintained our focus on the three-dimensional lifeworld. The time has now come for us to include the lower-dimensional lifeworlds in our account of cosmogony.

In general we have found that the lifeworld is:

1. *Quantized dimensional action.* It is not a static continuum, not a set of objective relationships contained in space, and not a subject that transcends the spatial container. It is the prereflective transpermeation of contained, container, and uncontained from which object-in-space-before-reflective-subject first arises.

2. *Cyclonic.* Lifeworld action is wave-like in nature. It has the quality of a spinning vortex.

3. *Organic.* The lifeworld changes, grows, undergoes dialectical development. It is transformed through several stages of Individuation.

4. *Polydimensional.* There are several lifeworlds.

5. *Topological.* The several lifeworld dimensions are embodied by the members of a topological hierarchy corresponding to the bisection series: Klein bottle, Moebius strip, lemniscate, sub-lemniscate.

With these lifeworld characteristics in mind, the task now at hand is that of describing the underlying patterns of cyclogenesis for all dimensionalities, the transformations of the several topodimensional vortices as they evolve in relation to each other. In this reckoning, we must also take into account the important asymmetry among the dimensions, as discussed in Chapter 5: lower-dimensional circulations entail less dialectical tension than do higher-dimensional ones. There is less differentiation and integration of subject and object in lower dimensionalities, less Individuation. In the case of the lowest dimensionality — the zero-dimensional sphere corresponding to the sub-lemniscatory member of the bisection series — there is no Individuation whatsoever. The zero-dimensional action that serves as carrier wave to higher-dimensional vortices can be said to be "selfless." It is the only action with nothing to call its own: no subject, no object, no dialectical tension, no space or time, no development. Generally then, the topodimensional interrelationships that we must describe are among circulations with differing degrees of dialecticity and developmental potential.

Consider again Table 5.2, the interrelational matrix of topodimensional spinors. Taken by itself, this table affords but a static picture of dimensional associations, one that is "averaged over," i.e. abstracted from, the actual facts of dimensional development. Therefore, to fill in the concrete details of how the several dimensional vortices evolve in relation to one another, we must set the matrix in motion. This is achieved in Table 7.1(a).

Table 7.1(a) displays the full course of development of all orders of topodimensional action. The unbracketed terms on the principal diagonal of each matrix (the diagonal extending from upper left to lower right) are our eigenvalues; they signify the relationship of a given dimension to itself (the terms enclosed in square brackets will be explained below). The ratios appearing in all other cells are members of enantiomorphic pairings that pertain to relationships *between* dimensional spinors (as will be discussed shortly).

The general design of Table 7.1(a) is circular. Moving upward from the matrix at the bottom, we have the clockwise, forward, or projective stages of dimensional generation, indexed by the values appearing to the left of the matrices. The turn to the counterclockwise, backward, or proprioceptive stages is then enacted by moving back down through those same matrices, with stage numbers now displayed to the right. The parenthetic terms accompanying each stage number indicate the topodimensional spinor or spinors to which that number applies; since the zero-dimensional spinor does not undergo development, it does not appear here. If the sequence of stages for each topodimensional eigenspinor is considered separately from that of the other spinors, we see that Table 7.1(a) in fact does not describe the action of a single circle but of circles nested within circles, so that the overall pattern is actually that of a *spiral*.

Clockwise Stages
(Projection)

Counterclockwise Stages
(Proprioception)

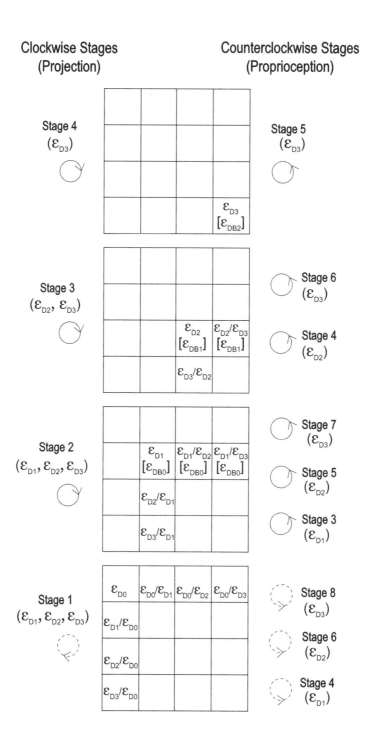

Table 7.1(a). Stages of dimensional cyclogenesis

To preserve the thoroughly interwoven, nonlinear character of dimensional interrelatedness, Table 7.1(a) displays the several windings of the dimensional spiral as overlapping one another. However, this makes the Table somewhat difficult to read. To facilitate understanding, I offer Table 7.1(b) as a visual aid. Here the circulations of the dimensional spiral have been parsed, teased apart for easier identification.

Upon inspecting Table 7.1(b), we observe the simple arithmetic increase in the number of matrices that are involved as we go from lower- to higher-dimensional windings of the dimensional spiral. This reflects the progressive increment in the number of stages through which cyclogenesis occurs. Note that something else changes in the process. From one winding of the spiral to another, there is a *logarithmic* increase in the number of cells found within each square matrix: 1, 4, 9, 16. This growth in matrix size signifies a progression in dimensional complexity, with larger matrices possessing higher degrees of dialecticity.

We saw in Chapter 5 that the strength of the dimensional dialectic, its intricacy and sharpness of definition, depend directly on the number of dimensional bounding elements that operate within the given lifeworld. These basic elements serve to differentiate and integrate the lifeworld. Thus, the three-dimensional Kleinian lifeworld, possessing three such bounding elements (point, line, and plane), is more dialectically refined than its lower-dimensional counterparts with their fewer bounding elements. In the matrices, the bounding or containing elements are distilled from interactions between the members of enantiomorphically paired cells that sit astride the principal diagonal. It is through the fusion of these enantiomorphs that lower-dimensional waves facilitate the formation of higher-dimensional waves. In each stage of its development, the higher-dimensional wave progresses by a fusion event in which a lower-dimensional "overtone" enantiomorph is introjected, absorbed into the higher-dimensional vortex in the interest of higher-dimensional self-containment, of differentiation and integration. The dimensional bounding elements distilled from enantiomorphic fusions are represented by the square-bracketed terms appearing within relevant cells of the matrices. We can see that larger matrices of Table 7.1(b) are able to accommodate greater numbers of different enantiomorphic pairings leading to the distillation of greater numbers of bounding elements, hence higher levels of dialecticity.

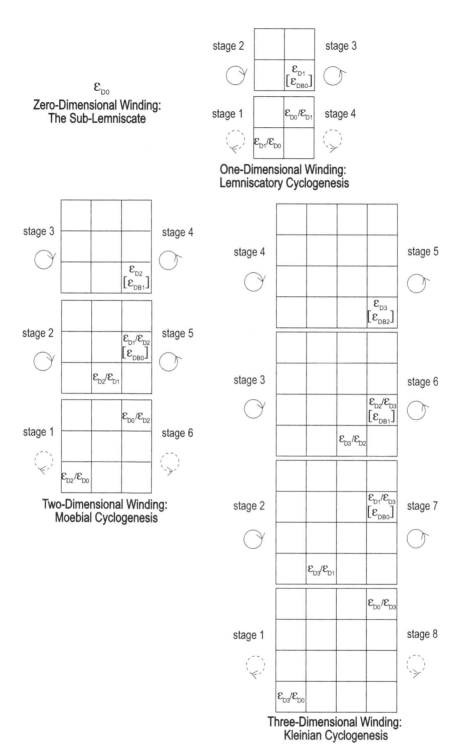

Table 7.1(b). Dimensional spiral, parsed into separate windings

(Before proceeding to examine in more detail the four windings of the dimensional spiral, let me add a comment on the nature of the bounding elements. Although I have associated them with points, lines, and planes, I am not primarily referring to the classical understanding of these elements in which they are assumed to be purely spatial, symmetric, structureless, and juxtapositional. While such an idealized view of points, lines, and planes does arise in the projective stages of development, the lifeworld reality that underlies this idealization is that the bounding elements, along with the dimensionalities they bound, are spatio-sub-objective and synsymmetric, featuring dialectical opposition, instead of being just structurelessly juxtapositional. The depth dimensional character of the bounding elements will be considered in due course.)

The first winding of the dimensional spiral is that of the sub-lemniscatory matrix, which consists of but a single cell. In effect, the "radius" of this circulation is "zero," for it entails no transformation whatsoever.[1] No stages of development can be found for ε_{D0} nor are any necessary, since this zero-dimensional action functions purely in the capacity of providing "selfless" guidance for the self-generation of the higher-dimensional actions.

Advancing to the second winding of the spiral, the matrix expands to the 2×2 structure associated with one-dimensional lemniscatory cyclogenesis. Again, while Table 7.1(b) displays this winding as separate from the others (enabling us to study it more easily), the fact is that no such simple detachment of circulations exists. But neither do the windings simply coincide, as they seem to do in Table 7.1(a). Rather, each developmental circulation possesses its own distinct timing and this precludes it from being merely simultaneous with any other.

The hands of a clock provide an elementary analogue of the different temporal "scales" associated with the different windings of the dimensional spiral. With a full cycle of the second hand, there is but a unit movement of the minute hand, and a full cycle of the minute hand corresponds to a unit movement of the hour hand. Thus we have faster cycles embedded within slower ones. Of course, in the lifeworld counterpart of this, the cycles are not simply commensurable. Lifeworld circulations do not constitute subdivisions of a well-established unitary time metric (comprising seconds, minutes, hours, days, etc.), but entail different degrees of temporality itself. As Heidegger noted, three-dimensional Being is the dimension of "true time" (1962/1972, p. 15); also, it is the dimension of *depth*, the *spatial* aspect of Being articulated by Merleau-Ponty. And, just as the lower dimensions possess less perspectival opposition thus less spatiality, they possess less temporality as well; the interplay of temporal elements (past, present, future) is weaker. For the very lowest dimension, there is no temporality or spatiality at all. The zero-dimensional sphere is one of absolute timelessness and spacelessness.

We are following the second winding of the dimensional spiral, the "fastest" circulation among those topodimensional actions that are subject to development. In its opening stage, we have the primordial matrix wherein the zero-dimensional sub-lemniscatory carrier wave, via the $\varepsilon_{D1}/\varepsilon_{D0}$ "overtone," supports or contains the fractionally one-dimensional lemniscatory "undertone," $\varepsilon_{D0}/\varepsilon_{D1}$. (It was established in Chapter 5 that, while each member of any given pair of dimensional enantiomorphs is a hybrid term spanning two different dimensions, the fractional member is an embryonic form of the higher dimension, whereas the native dimensionality of the reciprocal member is the lower dimension.) One-dimensional action is but nascently oriented here, as indicated by the dashed lines that compose the clockwise-directed circle; the possibility of forward projection is no more than embryonic.

Moving upward now to the stage-2 matrix of the lemniscatory winding, we find that embryonic or fractional one-dimensional action, $\varepsilon_{D0}/\varepsilon_{D1}$, has been brought to ε_{D1} integrality. This entails the quasi-fusion of enantiomorphs in which they are "annihilated," with the zero-dimensional enantiomorph, $\varepsilon_{D1}/\varepsilon_{D0}$, being absorbed into the one-dimensional lemniscatory vortex in such a way that the lower-dimensional containment or boundedness of stage 1 takes on the character of a point-based boundedness internal to the now-mature one-dimensional wave (in Table 7.1(b), internalization is signified by enclosure within square brackets). In this manner, the pointal or zero-dimensional bounding element is formed: ε_{DB0}. However, with the transition to stage 2, the primordial state of affairs is not simply left behind; rather, it is relegated to the background. The stage-2 occlusion of enantiomorphs is denoted by reduction of the matrix to but one active cell. The filling in of the clockwise circle of rotation indicates that oriented projective activity — the unidirectional movement *from* the lemniscatory subject *to* its object — is now occurring in earnest. Whereas the lemniscate's vortical action is but incipiently oriented in stage 1, it assumes a definite direction in stage 2 and the direction is "forward."

In the "clockwise forespin" of this phase, not only are the $\varepsilon_{D1}/\varepsilon_{D0}$ and $\varepsilon_{D0}/\varepsilon_{D1}$ enantiomorphs eclipsed, but also the ε_{D1} vortex itself, though the latter occlusion is not directly shown in the Table (the matrix reductions of Table 7.1 account only for the deeper occlusions of the lower-dimensional enantiomorphs). For this is the phase in which prereflective spatio-sub-objectivity (ε_{D1}) conceals itself in favor of the reflective subject. It is here that the one-dimensional lifeworld is repressed and the basic posture of projection is assumed, that of *object-in-space-before-subject*. However, I note again that, in this one-dimensional winding of the spiral, dialecticity is relatively weak. The lemniscatory subject is simpler, more rudimental than its higher-dimensional counterparts. One-dimensional differentiation and integration, one-dimensional Individuation, one-dimensional space and time, are all comparatively undeveloped. In moving to the Moebial and Kleinian windings of the dimensional spiral, dialecticity increases in the logarithmic fashion, with matrix size expanding from 2^2 to 3^2 to 4^2. But before said passage to higher dimensionality can take place, the forward spinning lemniscate must "switch gears," go into reverse so that the prereflective act of clarification that hitherto had been obscured can be proprioceptively acknowledged. It is by "going against the grain," by counter-acting the forward action of the lifeworld, that *awareness* is gained of this action, thereby bringing it to fruition. Quasi-Individuation then becomes genuine Individuation.

We know that the "reversal of gears" is embodied in the double cyclicity of topological action (symbolized most simply by the emblem of infinity, ∞). And what we find in Table 7.1 is a stagewise specification of the crucial transition from forward to backward action cycles for each order of cyclogenesis. Every action matrix is to be interpreted in two ways: initially as spinning forward or clockwise, subsequently as spinning backward or counterclockwise. With regard to the second winding of Table 7.1(b), we see that, in passing from stage 2 to stage 3, lemniscatory action has shifted in this way, the movement "against the grain" being expressed by the reversal of the arrowhead from clockwise to counterclockwise.

How does the Table elucidate the backward cycle of dimensional wave generation? It might be thought that retrograde lemniscatory action is only strongly manifested in stage 4, this stage being preceded by a stage-3 preliminary phase in which the "backspin" is merely incipient — just as the manifestation of stage-2 "forespin" is preceded by stage-1 nascency. It would appear then that, for the proprioceptive orientation to be fully realized in stage 4 of the lemniscatory winding, a period of gestation would be required in stage 3, as it was in stage 1 for the projective orientation of stage 2. However, this does not take into account that, in the second cycle, movement *proceeds in reverse*. The transition from stage 3 to stage 4 is not from embryonic action to well-oriented action, as in stage 1 to 2, but the other way around. In the proprioception carried out in stage 3, the decisively forward action of stage 2 that had obscured the lemniscatory lifeworld is counteracted by an equally strong retrograde action (indicated in Table 7.1 by the solidly drawn rotation circle); with this, lemniscatory action gains cognizance of itself. Still, this *is* only a first proprioception, one that does not include concrete apprehension of the lemniscate's primordial relation to the sub-lemniscatory carrier wave. It is in making the transition to stage 4 that the inchoate state of affairs prevailing at the outset of one-dimensional cyclogenesis is properly acknowledged. In stage 4, the embryonic clockwise movement of stage 1 is counteracted by an embryonic counterclockwise movement (denoted by the dashed rotation circle). With this proprioceptive return to the origin, the lemniscatory wave completes its formation in the "backward" fashion and the limited Individuations of stages 2 and 3 are surpassed. One-dimensional (w)holeness now reaches full-fledged concrescence.

To grasp the significance of stage 4, we must realize that while the stage-2 reduction of the 2×2 matrix is reversed here, this is no simple reversion to stage 1. For, unlike what happens in stage 1, the lemniscatory vortex is now moving *proprioceptively*. In this movement against the "grain" of stage 1, for the first time the dimensional organism gains concrete awareness of the stage-1 nascency from which its stage-2 projection originates. The matrix reduction of stage 2 signifies the fusion of enantiomorphs and concomitant introjection of the sub-lemniscatory carrier wave into the lemniscatory vortex. But this fusion in fact is a *quasi*-fusion. $\varepsilon_{D1}/\varepsilon_{D0}$ and $\varepsilon_{D0}/\varepsilon_{D1}$ enantiomorphs are actually not *completely* fused in stage 2, which means that the lower-dimensional carrier is not completely introjected. Instead of the lemniscate truly containing the sub-lemniscate, fully appropriating this lower-dimensional guiding wave for the purpose of its own development and thus gaining bona fide autonomy, the lemniscatory wave in fact remains dependent upon the sub-lemniscate, continues to be carried by it. So the zero-dimensional wave, operating "behind the scenes" in stage 2, maintains its support of the one-dimensional wave. Of course, ε_{D1} also operates covertly in stage 2, given that it has concealed itself in favor of the objectifying consciousness of the lemniscatory subject. In stage 3, the occlusion of ε_{D1} is counteracted by the proprioception that challenges the idealized autonomy of the one-dimensional subject; the spatio-sub-objective lemniscate thus gains a measure of self-awareness. Then, with the retrograde action of stage 4, the incomplete closure of ε_{D1} is acknowledged more fully by acknowledging in concrete fashion its continuing reliance on ε_{D0}. In so doing, ε_{D1}'s hidden dependence on ε_{D0} is brought to an end. But while the dependence is no longer obscured, does it not still exist? How then can we claim that ε_{D1} completes its development in this stage? The dialectical logic that governs dimensional development tells us that even though simple independence indeed is not achieved in the climactic stage, neither does simple dependence merely continue.

We can in fact say that the lemniscatory vessel becomes fully sealed in stage 4, with the sub-lemniscate being truly contained by it. But this happens by virtue of a dialectic whereby genuine closure entails *disclo*sure, or opening. The true containment of the sub-lemniscate involves lifting the repression of it that had been imposed with the matrix reduction of stage 2. In stage 4, the lemniscate contains the sub-lemniscate by acknowledging the latter's containment of *it*. We may conclude then that the containment realized in stage 4 is *mutual*: It is precisely when the zero-dimensional wave's containment of the one-dimensional wave is consciously recognized that the latter wholly contains the former!

Nevertheless, if we portray the containment of these dimensional vessels as mutual, we must be mindful of the *asymmetry* that is involved. We have seen that ε_{D0}'s act of containment is utterly "selfless," whereas ε_{D1}'s containment of ε_{D0} is carried out in the interest of completing its own development, its Individuation. Can we clarify further the sense in which ε_{D0}'s containment is "selfless"? It seems the term "selfless" must be taken quite literally. It is not that zero-dimensionality "generously gives of itself," but that it *has no* self. Then, if ε_{D0} is not a "giving presence" but an *absence*, its containment of the higher-dimensional wave evidently must be viewed in negative terms. Despite my earlier allusion to $\varepsilon_{D1}/\varepsilon_{D0}$ as ε_{D0}'s overtone enantiomorph, properly speaking ε_{D0} possesses no enantiomorph of *its own* with which to carry higher-dimensional action. Unable to enclose the higher-dimensional wave in any such positive fashion, the ε_{D0} carrier wave is limited to mediating the higher-dimensional wave's own appropriation of available enantiomorphic support in the interest of its self-containment. Since the enantiomorph in question thus actually belongs to the *higher*-dimensional vortex, the zero-dimensional wave can only "hold it in trust" for the one-dimensional wave, as it were. Therefore, when we speak of "mutual containment" in describing the lemniscatory winding's stage-4 relationship between zero- and one-dimensional waves, it must be understood that the carrier wave, for its part, engages in a *negative* kind of containment: it contains the one-dimensional wave only in the sense of mediating its *self*-containment. The elusiveness of this notion cannot be helped since what we are attempting to talk about here in considering the "selflessness" of the carrier wave is an enigma that confounds language. In such an effort, we must rely exclusively on metaphor.

Does stage 4 close the circle of development? It closes the lemniscatory winding of the dimensional *spiral*, which means that there is also an opening into a new winding. In completing the round of 2 × 2 matrix generation, logarithmic expansion carries us into the inaugural phase of the 3 × 3 round. The compass has now shifted so that action is once again nascently clockwise. In stage 1 of the two-dimensional Moebius winding, the carrying capacity of the zero-dimensional sub-lemniscate has expanded. Through the new enantiomorph, $\varepsilon_{D2}/\varepsilon_{D0}$, the sub-lemniscate offers supportive containment (in its negative way) for the embryonic $\varepsilon_{D0}/\varepsilon_{D2}$ vortex. Then, with the fusion of these enantiomorphs that reduces the matrix and brings us to stage 2, a quasi-closure (and opening) is achieved; the sub-lemniscatory wave is absorbed into the Moebius wave in such a way that zero-dimensional support ($\varepsilon_{D2}/\varepsilon_{D0}$) becomes two-dimensional self-support via distillation of the zero-dimensional bounding element (ε_{DB0}). In this stage, we have a first projection of object-in-space on behalf of emergent two-dimensional subjectivity. However, whereas the one-dimensional lemniscatory wave required but a single projection to complete its quasi-closure, the more complex Moebius pattern of development calls for a second such projection to achieve integrality — as we can see from the still fractional form $(\varepsilon_{D1}/\varepsilon_{D2})$ assumed by the Moebius wave in stage 2.

In the Moebius winding of the dimensional spiral, the lemniscate has attained the status of carrier wave and so can support the further development of the Moebius. The lemniscate does this through a feature that it does not possess in its own cycle of generation. It now has a dimensional overtone, $\varepsilon_{D2}/\varepsilon_{D1}$, which, in stage 2, is enantiomorphically coupled with the Moebius undertone, $\varepsilon_{D1}/\varepsilon_{D2}$. It is the fusion of these enantiomorphs that brings us into the third stage of Moebius cyclogenesis, the conclusive stage in its forward-directed autopoiesis. A second quasi-closure occurs here and the lemniscatory carrier wave dissolves into the now integral Moebius wave, ε_{D2}, with a concomitant distillation of the one-dimensional bounding element, ε_{DB1}. This is the second and final projection of object-in-space on behalf of the two-dimensional subject.

Next comes the return to the Moebius lifeworld via proprioception. Entering stage 4, the "gears" are put into reverse and two-dimensional spin is reoriented. In this first proprioception, the divisive posture of stage 3 is challenged by an initial self-awareness of Moebial (w)holeness. Then, moving concretely backward in stage 5, two-dimensional proprioception is carried further. The Moebius organism now acknowledges the "selfless" support of the lemniscatory carrier wave; in the process, the asymmetrically mutual containment of Moebius and lemniscatory waves is brought about. Going still further back in stage 6, the contribution of the sub-lemniscatory carrier wave is recognized by the Moebius wave so that *these* two spheres of dimensional action become harmonically interrelated. The 3 × 3 matrix reduction has now been wholly counteracted and the Moebius vortex has completed its formation in earnest. Its Individuation has fully been achieved.

Following the proprioceptive closure of the 3 × 3 Moebius circulation, the dimensional spiral dilates once again, the compass reverts to clockwise, and we find ourselves in the opening phase of the 4 × 4 epoch of dimensional vortex generation. Beyond its incipiently oriented stage, the three-dimensional Kleinian organism develops through three clockwise phases of projection. Three quasi-closures occur, three introjections of lower-dimensional carriers, accompanied by distillations of corresponding bounding elements. In stage 2, $\varepsilon_{D3}/\varepsilon_{D0}$ and $\varepsilon_{D0}/\varepsilon_{D3}$ enantiomorphs fuse to produce the still fractional three-dimensional wave, $\varepsilon_{D1}/\varepsilon_{D3}$, whose ε_{DB0} bounding element is distilled from the introjection of the sub-lemniscatory $\varepsilon_{D3}/\varepsilon_{D0}$ support wave. In stage 3, the Kleinian vortex advances to $\varepsilon_{D2}/\varepsilon_{D3}$ via the merger of $\varepsilon_{D3}/\varepsilon_{D1}$ and $\varepsilon_{D1}/\varepsilon_{D3}$ enantiomorphs, with the ε_{DB1} bounding element of $\varepsilon_{D2}/\varepsilon_{D3}$ crystallizing from the introjection of the $\varepsilon_{D3}/\varepsilon_{D1}$ support wave. Finally, in stage 4, $\varepsilon_{D3}/\varepsilon_{D2}$ and $\varepsilon_{D2}/\varepsilon_{D3}$ enantiomorphs fuse to bring the three-dimensional wave to integrality, with the ε_{DB2} bounding element of this ε_{D3} vortex being distilled from the introjection of the $\varepsilon_{D3}/\varepsilon_{D2}$ carrier wave. Three projections of object-in-space on behalf of the three-dimensional subject thus take place in the comparatively complex quasi-Individuation of Kleinian dimensionality.

Subsequently, "gears are reversed" for a return to the lifeworld that will bring genuine Individuation. This unfolds over a series of four proprioceptions. In stage 5, the (w)holistic Kleinian source of the trichotomous world projected in stage 4 comes to light. Here the dimensional organism recognizes the pre-reflective ground of the reflective subject and its spatially framed objectifications. Then, in the proprioceptions of the three ensuing stages, cyclonic Individuation progressively advances via the Kleinian wave's acknowledgments of the support it has received from the three lower-dimensional vortices. And each acknowledgment brings about the asymmetric mutual containment of the Kleinian wave and a lower-dimensional wave. When all acknowledgments have been given — both in the Kleinian winding of the dimensional spiral and in the other topodimensional windings that overlap it — the four spheres of dimensional spin contain one another harmoniously.[2]

In general, the foregoing analysis brings to light the substages of dimensional development that serve to distinguish one topodimensional spin structure from another. While the several windings of the dimensional spiral do overlap, each lifeworld circulation runs its own course, with distinct circulations being marked by differences in the number of substages that each requires to carry out its clockwise projections and counterclockwise proprioceptions, along with differences in the sizes of the dimensional action matrices corresponding to the stages. These differences reflect, in turn, differing degrees of dialecticity, of lifeworld complexity, of the capacity for Individuation. Having now completed our theoretical study of how dimensions evolve, we can apply the principles we have discovered to the issue of primary concern to us: the co-evolution of nature's fundamental forces.

Meta-mathematical postscript

Before proceeding to the next chapter for the promised application, let me acknowledge that, by and large, I have described dimensional generation in an intuitive and qualitative way. A more detailed algebraic account could be provided. For example, "annihilation operators" could be introduced that would furnish mathematical details about the process of enantiomorphic fusion and the accompanying distillation of dimensional bounding elements. The problem with implementing such an approach is that — while dimensional development might thus be clarified in abstract, mathematically objective terms — what we presently seem to require is a different kind of clarity.

The prevailing approach to mathematical physics is guided by the aim of achieving objective knowledge of nature via formal precision. The mathematical structures in play are pure abstractions, universal objects that, once arrived at, must then be put into correspondence with the concrete particulars of empirical observation. We know that this task has proven to be extremely difficult for contemporary unification physics. The upshot of what we have found on this in previous chapters is that the underlying phenomena of modern physics in fact do not lend themselves to being treated as objects-in-abstract-space appearing before a detached mathematical subject. What I have proposed is that the phenomena in question be seen as *spatio-sub-objectivities* possessing a distinctly concrete aspect. The quantized spin structures of contemporary physics are neither abstract universals nor concrete particulars but *concrete universals*. Merleau-Ponty afforded a glimmer of concrete universality in his description of the "flesh of the world," a concept closely akin to that of depth:

The flesh is not matter, in the sense of corpuscles of being...is not mind, is not substance. To designate it, we should need the old term "element," in the sense it was used to speak of water, air, earth, and fire, that is, in the sense of a *general thing*, midway between the spatio-temporal individual and the idea, a sort of incarnate principle that brings a style of being wherever there is a fragment of being. The flesh is in this sense an "element" of Being. (1968, p. 139)

The lifeworld dimensions we have been considering are of this kind. They are *general things*, not mere generalities or particularities. Formalist mathematical physics — rooted as it is in an unacknowledged intuition that implicitly divides the general and the particular, the abstract and the concrete — cannot effectively encompass the concrete universality of the lifeworld. What we require is *phenomenological* intuition, a faculty that seems better served by geometric imagination than by the formal abstractions of algebra. I submit then that the topological structures we have been working with (the Klein bottle, the Moebius strip, etc.), by virtue of their concrete dialectical nature, can effectively facilitate the phenomenological intuition necessary for directly apprehending the spatio-sub-objective lifeworld, whereas that world is unapproachable via the abstract formalisms of extant mathematical physics. Therefore, qualitative geometric intuition will continue as the method of choice in the pages that follow. (For more on the new kind of clarity provided by the phenomenological approach to contemporary physics, see Chapter 9.)

Notes

1. Of course, the "radii" of *all* dimensional circulations are "zero," in the sense that stage transitions are not actually displacements in an extensive spatial continuum (though they appear as such in Table 7.1) but rather — in the language of Heidegger — "*pre*spatial" gestures of "opening up" that first "provide space" (1962/1972, p. 14; emphasis added). In other words, the several dimensional circulations portrayed in the Table cannot really have finite radii because — instead of taking place within a spatial field of reflection, they constitute the prereflective actions from which the several fields of reflection derive.

2. If dimensional process is indeed most essentially described not by a circle but by a *spiral*, we should expect that the account of cyclogenesis would not end at this point. Instead there would be another logarithmic expansion that would mark the inauguration of a new round of dimensional autopoiesis, one characterized by the generation of the 5 × 5 matrix of a four-dimensional vortex. That is, a new order of the lifeworld would grow organically. This idea is adumbrated further in Chapter 10.

Chapter 8

The Forces of Nature

8.1 The Phenomenon of Light

In Chapter 5, I described the four topodimensional eigenwaves of Table 5.2 as primary forms of movement. I am now ready to take up the idea foreshadowed at the beginning of Chapter 6 that these forms of movement or wave motion are none other than the field motions associated with the four fundamental forces: electromagnetism, the strong and weak nuclear forces, and gravitation. From the microphysical standpoint, the smallest units of the primary force fields are subatomic particles known to physicists as *gauge bosons*. The electromagnetic force is associated with the photon, the weak nuclear force with the weak gauge boson, the strong nuclear force with the gluon, and the gravitational force with the hypothesized graviton. I am proposing then that the basic force fields and their associated gauge bosons can be put into correspondence with the eigenvalues of our matrices (Tables 5.2 and 7.1), giving specific physical meaning to those values. Since the work of Chapter 7 has enabled us to see how eigenwaves overlap and synchronously interweave in the course of their development, associating these waves with the forces of nature brings the latter into dynamic harmony.

However, have I not gone so far as to suggest that the eigenspinors constitute sub-objective dimensions unto themselves, whole lifeworlds? Can the same be said of the forces of nature? To show that this is so, let us begin with the electromagnetic force.

In founding quantum mechanics, Planck dealt with but one force of nature: electromagnetism. Planck, like Michelson and Morley before him, confronted the implicit discontinuity associated with electromagnetic action. The Michelson-Morley experiment of 1887 had called into question Maxwell's presupposition of a continuous (ethereal) medium for the propagation of light (we have seen how Einstein subsequently preserved the notion of continuity by replacing the classical continuum with the more abstract space-time continuum). In a very different experiment conducted at the end of the following decade, Planck also encountered discontinuity in working with electromagnetism (as discussed in Chapter 2). His blackbody research led him to the inescapable conclusion that electromagnetic energy is quantized, that it comes in discontinuous bundles of action, which later came to be called photons. It should be clear by now that discontinuity is abhorrent to classical intuition. The aim of objectifying nature critically depends on the availability of a continuous medium or context in which the objectification can take place. But the phenomenon of light uniquely resists such objectification. Why is this so?

It is an obvious fact of ordinary perception that the visual appearance of an object changes concretely in accordance with changes in the perspective from which it is viewed (unless the object is perfectly symmetrical, like a sphere). In shifting the position of my head, for example, the objects on my desk (computer, books, telephone, etc.) are exhibited to me somewhat differently owing to the modified angle of view, with surfaces that had been concealed now becoming visible, and vice versa. The visual perception of objects naturally involves the reflection of light. What happens when the "object" being considered is light itself? When Michelson and Morley measured the velocity of light from different frames of reference, instead of encountering the differences that they and every other researcher had fully expected, light's velocity remained the same. In the perceptual analogy, it would be as if the objects on my desk would look exactly the same to me regardless of my angle of view. Michelson and Morley's strange discovery was most alarming to physicists. In fact it was a bombshell that was soon to precipitate the Einsteinian revolution. That is because the finding of Michelson and Morley did nothing less than call into question the classical intuition of object-in-space-before-subject that had implicitly governed human experience for many centuries.

The maintenance of classical identity is essentially predicated on separating subject and object in such a way that, whereas objects change concretely, the subject and the space in which it operates remain abstractly the same. The concrete change I expect to see when I shift my position with respect to the objects I perceive presupposes that the spatial context within which I make this observation itself does *not* change, that its continuity is preserved. By the same token, the presupposition is that *I* do not change, that the "invisible I" behind these eyes remains in-*di*-visible, retains its individuality. Because a solitary individual lies behind the multiplicity of concretely shifting perspectives, said perspectives are readily interchangeable, with one being continuously transformable into another in a predictable fashion. Confirmation of these predictions serves to confirm the *cogito's* sense of himself. When, upon altering my perspective, I see how the appearance of an object I have been viewing changes as expected, I am indeed reassured that I have *not* changed. What Michelson and Morley saw was not reassuring.

Just why was it that the velocity of light did not change regardless of the frame of reference these researchers adopted? Why did light "look" the same to them no matter what perspective upon it they assumed? I propose it was because, in confronting the phenomenon of light, they were not encountering an object to be seen, but that *by which* they saw. In this regard, cosmologist Arthur Young (1976) comments that while the conventional "scientist…likes to think of [a photon of] light as 'just another kind of particle,'…light is not an objective thing that can be investigated as can an ordinary object….Light is not seen; it is [the] seeing" (p. 11). The physicist Mendel Sachs reaches a similar conclusion in his inquiry into the meaning of light: "What is 'it' that propagates from an emitter of light, such as the sun, to an absorber of light, such as one's eye? Is 'it' truly a thing on its own, or is it a manifestation of the coupling of an emitter to an absorber?" (1999, p. 14). Sachs's rhetorical question suggests that the phenomenon of light — instead of lending itself to being treated as a changeable object open to the scrutiny of a changeless subject that stands apart from it — may be better understood as entailing the *inseparable blending of subject and object, of changelessness and change.*

While the objects on my desk do not look the same to me when I view them from different perspectives, in attempting to observe the light by which these objects are perceived, it seems I would be confronted with the self-referential prospect of "viewing my own viewing," and this would mean that I would not encounter the concrete variations in appearance that attend the observation of an object from a fixed viewpoint that itself is not viewed. But would my encounter with light level all differences in perspective and convey an impression of simple sameness? This would be the result only if I continued to maintain the objectifying posture of the detached subject, as Michelson and Morley evidently did. On the other hand, if, instead of treating light as an object, I were actually to *enter into* it via phenomenological intuition, then I would not experience light simply as something that looks the same from all perspectives, but would experience the dialectic of sameness and difference that accompanies the blending of subject and object.

The phenomenon of light indeed plays a central role in phenomenological investigation. We might say that Merleau-Ponty's study (in "Eye and Mind") of Cézanne's "autofigurative" art essentially studies the paradox of light (see Chapter 3). Heidegger, for his part, is quite explicit about the importance of light to phenomenological thought. After acknowledging the contributions of Hegel and Husserl in surpassing the old mechanistic objectivism and making subjectivity "the matter of philosophy" (1964/1977, p. 383), Heidegger comments that — in this thinking of subjectivity into its own, "to its ultimate originary givenness,…to its own presence" (p. 383), something remains *un*thought:

> What remains unthought in the matter of philosophy as well as in its method? [Hegel's] speculative dialectic is a mode in which the matter of philosophy [i.e. subjectivity] comes to appear of itself and for itself, and thus becomes present. Such appearance necessarily occurs in some *light*. Only by virtue of light, i.e., through brightness, can what shines show itself, that is, radiate. (p. 383; italics mine)

Evidently then, what remains unthought in the history of philosophy is the phenomenon of light, or what Heidegger later calls *enargeia* ("that which in itself and of itself radiates and brings itself to light"; p. 385).

It is clear that, for Heidegger, *enargeia* or light is not merely a local, objectively observable phenomenon, not just a finite particular being. Heidegger implicitly associates light with *Being*, with "presence as such," rather than just with "what is present" (1964/1977, p. 390). And, according to philosopher Carol Bigwood, while Heideggerian "Being is not a being," neither is it a "God [or] an absolute unconditional ground…but is simply the living web within which all relations emerge" (1993, p. 3). That is to say, *Be-ing* constitutes the dimension of dynamic life process, the lifeworld dimension. From this we can conclude that light, or, more generally, electromagnetism, indeed comprises a sub-objective dimension unto itself, an entire lifeworld. Thus, light as such (as opposed to that which merely is lit), light as quantized action ($\hbar$), is the paradoxical phenomenon that gives physical significance to Merleau-Ponty's dimension of depth. And since the depth dimension has been fleshed out topologically via the Klein bottle (Chapter 4), we can say that the sub-objective, self-containing Kleinian vortex is a vortex of light. That is, the electromagnetic force can be associated with ε_{D3}, the Kleinian eigenvortex of Tables 5.2 and 7.1. If it is asked why electromagnetic action should specifically be identified with the Kleinian vortex and not with one of the other three eigenwaves we have identified, the answer is that only the three-dimensional Kleinian member of the topodimensional series can accommodate the three-dimensional dialecticity of light: it is through the exchange of photons that we directly experience our three-dimensional universe.

What of the sub-Kleinian members of the series? The proposition I am offering is that each of these lower-dimensional vortices has its unique gauge bosonic counterpart. More specifically, I suggest that the ε_{D2} Moebius wave is associated with the weak gauge boson, the ε_{D1} lemniscatory wave with the gluon, and the ε_{D0} sub-lemniscatory wave with the graviton. In preparing to examine this proposal, let us return to our consideration of Kaluza-Klein theory.

8.2 Phenomenological Kaluza-Klein Theory

To reiterate, Kaluza-Klein theory is based on the general idea that, by adding dimensions to the description of nature, we increase the capacity for incorporating additional force fields and their associated gauge bosonic particles. This strategy derives from Kaluza's original discovery that expressing Einstein's gravitational theory in five dimensions yields Maxwell's equations for electromagnetism. The goal of contemporary Kaluza-Klein theory is to encompass all four forces of nature in a single reckoning by taking into account six dimensions presumed to be microphysically concealed within the three dimensions of space and dimension of time comprising the known universe. The forces are thus unified in ten dimensions, six of these being curled up at the Planck length (see below for a discussion of the eleven-dimensional version of string theory given in M-theory).

We examined the problem with extant Kaluza-Klein theory in Chapter 4. Putting it in the philosophical language of the previous chapter, what we have is an unbridgeable gap between the general and the particular. On the one hand, there are the abstract generalities of the mathematical analyst's equations and his or her epistemological space. On the other hand, there are the particularities of the objectified topological spaces and the string-like, Planck-scaled particles that vibrate within them. It is because the curled up six-dimensional space is treated as an object that it possesses the ontological status of a finite particular being whose characteristics are open to indefinite variation. The topological space could take on a great many different shapes within the enormous class of Calabi-Yau spaces. Since direct observation is impossible and the theory offers no guiding principle that could narrow the range of possibilities, theorists have found it extremely difficult to arrive at a specific prediction of the structure of the six hidden dimensions that could then be tested against what experiments already tell us about the properties of the forces of nature. Because achieving a unified account of the four forces depends on specifying the precise structure of the dimensions in which the force particles vibrate, unification has indeed been elusive.

The alternative approach I am proposing challenges classical intuition. Here dimensions are not regarded as particular objects cast before the analytic gaze of the detached mathematical subject, but as spatio-sub-objective concrete universals whose dynamically evolving structures are directly accessible to phenomenological intuition. What is the role of the force particles in this?

We have seen from Merleau-Ponty that the depth dimension is not a spatial framework devoid of content, nor are the contents simply separate from that which contains them (as are objects in a box, for example). Instead depth dimensionality is *self-containing* (Chapter 3). This is how we are to think of the gauge bosons. Rather than being objects contained in space, these string-like particles are the self-containing, cyclonically vibrating spatio-sub-objectivities themselves; as with the photon, each of the force particles is a lifeworld dimensionality unto itself (a "global 'locality'"; Merleau-Ponty 1964, p. 180). So the four types of gauge bosons[1] constituting the primary forces of nature correspond to our four topodimensional eigenspinors. Since the characteristics of these quantized action structures are specifically correlated with the concrete topological properties of the members of the bisection series, the need for elaborate guessing games regarding the exact "shape" of the hidden dimensions is thus obviated.

We must bear in mind, however, that while observing the sub-Kleinian members of the bisection series as surfaces in three-dimensional space provides us with a distinct indication of the structure of the lower dimensions, it clearly falls short of achieving full insight into those dimensions. The spatio-sub-objective qualities of the Moebial, lemniscatory, and sub-lemniscatory structures are indeed obscured when they are regarded as mere objects in three-dimensional space. We have found that, through phenomenological intuition, we can gain an immediate sense of the sub-objectivity of our Kleinian lifeworld. To do the same for lower-dimensional lifeworlds requires orders of sensibility that are different from the one that operates three-dimensionally. Different lifeworlds necessitate different modes of subjective functioning for their direct apprehension. What we are seeing is that the nature of subjectivity matters in the phenomenological approach. It seems then that a full-fledged phenomenological physics in fact must be thoroughly *psycho*physical; a psychological aspect is indispensable to it. This issue will be considered in the closing chapters of the book.

The account I have given of the co-evolving family of topodimensional spin structures in fact constitutes a preliminary phenomenological rendering of ten-dimensional string theory, with its four observable dimensions (three-dimensional space plus time) and its six "compact" ones. How do the dimensions of phenomenological string theory add up to ten? Overall ten-dimensionality is not obtained by formalistically invoking six extra dimensions that go beyond the 3+1-dimensionality of the known universe, but by the more parsimonious procedure of considering only the dimensional elements nested *within* the 3+1-dimensional world, those corresponding to the basic entities intuited by Poincaré in founding dimension theory: the zero-dimensional point, the one-dimensional line, and the two-dimensional plane. In our topo-phenomenological rendition, these entities are *re*intuited via the sub-lemniscate, lemniscate, and Moebius, respectively. Moreover, instead of limiting the three terms in question to the role of ready-made bounding elements of a singular three-dimensional world, each is seen to be associated with a spatiotemporal world unto itself, with each evolving in distinct epochs of generation (with the exception of the zero-dimensional world, which is utterly spaceless and timeless, and lacks evolutional potential). The dimensional scheme thus yielded can be expressed as follows: $(0+1) + (1+1) + (2+1) + (3+1) = 10$. Within each bracketed term, the left-hand member is the spatial component and the right-hand member is the added component of time. Summation over the first three terms gives the six hidden dimensions. Note that, in conventional string theory, the number of dimensions in which the equations are written is determined wholly on practical grounds: only when the equations are cast in ten dimensions can nonsensical negative probabilities be eliminated from the calculations. This purely formal criterion for establishing the number of requisite dimensions does not help us gain an intuitively satisfying understanding of why ten dimensions are needed.

However, without unpacking an important distinction, the subtlety of phenomenological string theory is lost to oversimplification. It is not the case that the four gauge bosonic eigenspinors of our spin matrices (Tables 5.2 and 7.1) are simply identical to the four space-time terms in the ten-dimensional formula. For we have discovered that our eigenspinors are *pre*-spatiotemporal; they are the prereflective spatio-sub-objective actions that first open up the manifest space and time in which reflective action is possible. So it is not so much that the Kleinian eigenspinor, ε_{D3}, *is* 3+1-dimensional space-time, but that said spinor *projects* this space-time structure. Similarly, ε_{D2} projects 2+1-dimensional space-time, ε_{D1} projects 1+1-dimensional space-time, and ε_{D0} apparently projects 0+1-dimensional space-time.

Of course, the differences among the eigenspinors are not merely quantitative or even qualitative; they are *ontological*. Each prereflective dimensional action expresses a unique lifeworld, and spins out ontologically unique manifestations of space and time. In the differently dimensioned cyclonic projections of object-in-space-and-time-before-subject, there arise *distinct forms* of subject and object, and of space and time. Needless to say, this is not reflected in the formula for ten-dimensionality written above. Moreover, given the uniformly additive relationship between spatial and temporal terms shown in the formula, we find no hint of the fact that space and time become *less differentiated* from each other as we descend into the weaker dialectical oppositions of the lower dimensions. Just as subject and object are less differentiated from one another the lower we go, so too with space and time, until we reach the zero-dimensional sphere of utter space-timelessness. The formula's "0+1" term is especially misleading in this regard, for hardly do we have a single positive dimension of time in this most primal realm. Instead, the sub-lemniscatory wave constitutes a *null* dimension, a dimension of sheer nothingness, as it were.

With regard to the pre-spatiotemporality of the eigenspinors in phenomenological string theory, recall the discussion of spin and vorticity in Chapter 4. There we examined theories that highlight microphysical spin as the primordial action from which space, time, and the phenomena of quantum mechanics are first generated. One approach mentioned was Penrose's concept of the "twistor." In the late 1970s, this theory was widely viewed as an important contender in the effort to account for quantum gravity. However, over the next twenty years, twistor theory was overshadowed by the ascendancy of superstring theory. During that period, relatively little heed was paid to the idea of pre-spatiotemporal spin. But the picture is currently changing. Ed Witten (2004), the premier theorist of strings, is now leading the effort to formulate a "twistor-string theory" that seeks to marry string and twistor concepts by embedding string theory in twistor space. This opens the way for establishing pre-spatiotemporal spin as the key idea of quantum gravity.

For my part, I am proposing that a detailed account of the elementary forces of string theory may be afforded by embedding the theory in the pre-spatiotemporal matrix of spin structures initially set forth in Chapter 5. The matrix in question constitutes a special application of the Clifford-based hypernumber idea (Chapter 4) that provides a highly specific rendition of primordial vortex action ("twist"), a topodimensional array of four eigenvortices directly associable with the four types of gauge bosons found in nature. I am also suggesting, of course, that the harmony of forces can be brought to complete fruition only by including in our account the phenomenological roots of pre-spatiotemporal spin, rather than by seeking to objectify said cyclonic action in adherence to the old formula of object-in-space-before-subject. In fact, a measure of support for the concretely universal character of bosonic strings may be found in speculations arising from the recent refinement and generalization of superstring theory known as *M-theory*.

There is a deficiency in conventional string theory that I have not mentioned before now. The ambiguities of the theory are compounded by the fact that it actually comes in five distinct versions, with no basis for choosing which of the five renditions would uniquely correspond to the facts of the physical world. Beginning with the work of Witten (1995) and Horava and Witten (1996) that probed beyond the Planck-scaled realm to finer levels of analysis, the problem was addressed by constructing a kind of master theory — M-theory — in which equivalences ("dualities") among the different versions of string theory could be demonstrated. An important implication of M-theory relevant to our immediate concern is that the basic ingredients of nature are not limited to one-dimensional strings, but might include higher-dimensional elements called "branes" (as in "membranes"). In this scheme, we may think in terms of "two-branes" (two-dimensional membranes), "three-branes," or more. Greene speculated that our entire universe might be conceived as a brane:

> If we are living within a three-brane — if our four-dimensional spacetime is nothing but the history swept out by a three-brane through time — then the venerable question of whether spacetime is a something would be cast in a brilliant new light. Familiar four-dimensional spacetime would arise from a real physical entity in string/M-theory, a three-brane, not from some vague or abstract idea. In this approach, the reality of our four-dimensional spacetime would be on a par with the reality of an electron or a quark. (2004, p. 391)

So M-theory may provide an answer to "the centuries-old question regarding the corporeality of three-dimensional space" (Greene 2004, p. 412). What does this mean from a philosophical standpoint?

Greene is conjecturing that space-time, conceived as a "braneworld" (p. 386), is not a "vague or abstract idea" but "a something." Is this notion not reminiscent of Merleau-Ponty's view of the lifeworld as a *general thing*" (see previous chapter)? Classically, there is a peremptory distinction between space and time, on the one hand, and the objects contained therein, on the other. Space and time are not themselves tangible things but comprise the abstract framework within which concrete things are observed and manipulated. But have we not found that, beginning with Einstein, space and time have indeed been turned into objects? The six-dimensional Calabi-Yau space of string theory is thus a mathematical object embedded in a ten-dimensional analytic continuum. And yet, while Calabi-Yau space may implicitly be treated as an abstract object in higher-dimensional space, with respect to the one-dimensional strings it contains Calabi-Yau is regarded strictly as a space. In both cases, the sharp division is maintained between the containing space and the objects embedded therein. Does M-theory intimate a transcendence of this classical dichotomy?

The appearance of branes in M-theory is linked to the emergence of a new, eleventh dimension beyond the ten of conventional superstring theory. Greene notes that the ten-dimensionality of the older approach is determined by "counting the number of independent directions in which a string can vibrate, and requiring that this number be just right to ensure that quantum-mechanical probabilities have sensible values." In M-theory, however, "The new dimension…is *not* one in which a…string can vibrate, since it is a dimension that is locked within the structure of the 'strings' themselves" (1999, p. 310). The idea of a higher-dimensional space "locked within" its lower-dimensional objects certainly seems odd. Indeed, it runs counter to the classical intuition of objects in space. Evidently, M-theory's eleventh dimension does not simply contain its two-dimensional brane — not in the familiar way in which a container exists independently of its contents. Nor would it be any more accurate to say that the two-brane contains the eleventh dimension in this way. How then may we understand the containment relation implicit in M-theory?

I propose that the dimensionality of M-theory bespeaks the relation of *self*-containment characteristic of the *depth* dimension. Recall that Merleau-Ponty, in reflecting on the work of Cézanne, asserted that "we must seek space and its content *as* together." This seems applicable to M-theory's "eleventh" dimension and the branes that are "contained" by it. The brane is not merely a particular thing but a *general* thing, a concrete universal, for the brane and its spatial container mediate each other internally; they flow together in an unbroken circulation. It is in this sense that the space-time associated with the braneworld "is a something"; that it is not just a "vague or abstract idea" (Greene 2004, p. 391) but a corporeal presence. (The brackets I have placed around the term "eleventh" indicate the misleading nature of this term. For it seems clear that the dimension uncovered by M-theory does not extend the ten exterior dimensions of conventional string theory in a merely quantitative way by adding one dimension of the same kind; rather, M-theory introduces a *different* kind of dimensionality. By the same token, the brane does not simply extend the one-dimensionality of the object in space to two dimensions, since, unlike the conventional string, the brane surpasses the status of object by encompassing space itself.)

From what we found earlier, we might expect such radically non-classical relations to arise in the ultra-microscopic realm below the Planck length. Here space-time as ordinarily conceived, and all the presuppositions that accompany it, lose their coherence and applicability. In M-theory's high-resolution refinement of conventional string theory, the initial assumption was that the "eleventh" dimension is indeed scaled considerably below the Planck length. But wouldn't a brane*world* have to be much *larger* than the Planck length? In fact, would it not encompass all of space-time? Despite their original assumption, string/M-theorists do feel free to explore the possibility of a much larger "eleven" dimensional regime and correspondingly larger branes. How is this exploration justified? It is permitted by the fact that the equations of string/M-theory actually cannot specify the size of the extra dimensions, any better than they can specify their shape. Evidently, while the general equations of the theory are expressly designed to "tame" the ambiguities of the trans-Planckian world, ambiguity is carried forward epistemically, making its presence felt in the multiplicity of possible solutions to those equations, with no provision of a guiding principle by means of which the field of possibilities can be reduced.

For my part, I suggested in Chapter 2 that the uncertainty about scale most deeply reflects the fact that, "below" the Planck length, the linear scale of magnitude collapses, thereby confounding the very concepts of "large" and "small." Approaching this phenomenologically, we do not take ambiguity simply as a mark of failure and look to suppress it. Instead we accept it constructively, since phenomenological intuition is not limited to the "unambiguous" logic of either/or, but grounds itself in dialectical integration. The scale uncertainties of the trans-Planckian realm bring to mind once again Merleau-Ponty's characterization of the depth dimension as a "global 'locality'" in which we cannot sharply divide the "local" elements of space from space as a whole. Seen in this light, the "sub-microscopically scaled" and "cosmically scaled" branes can be regarded as the self-same scale-defying brane. (Is the scale-ambiguous brane not reminiscent of the *black hole* [see Chapter 2]? Greene confirms the scale ambiguity of black holes [2004, p. 338], having disclosed earlier [1999, p. 330] that branes can actually assume the appearance of black holes!)

Now, if the braneworld scenario is coupled with Witten's (2004) proposal of embedding string theory in Penrose's "twistor space," and if my phenomenological rendition of "twistor-string theory" is brought to bear on the copula, braneworlds become identified with cyclonically spinning *lifeworlds*. In particular, the four differently dimensioned eigenvortices of my topodimensional matrices become associated with four hierarchically ordered branes. On this account, branes are not exterior space-time worlds per se, but are the pre-spatiotemporal actions that first spin out those worlds: the Kleinian brane gives the visible 3+1-dimensional universe, the Moebial brane gives 2+1-dimensional space-time, the lemniscatory brane gives 1+1-dimensional space-time, and the sub-lemniscatory brane gives "0+1"-dimensional "space-time" (the latter, in fact, is a *null* dimension devoid of space and time, as discussed above). Thus, in terms of the external space-time dimensions that are generated, the dimensions still add up to ten in number, with M-theory's "eleventh" dimension constituting the interior or process dimension, the depth-dimensional braneworld action that does the generating. And each primordial action corresponds to a different force of nature. To reiterate my proposition: Kleinian spin is associated with electromagnetism, Moebial spin with the weak nuclear force, lemniscatory spin with the strong nuclear force, and sub-lemniscatory spin with gravitation. In the next chapter, this hypothesis will gain support in turning to the question of cosmogony. Let me close the present chapter by summarizing some basic differences between the conventional and phenomenological approaches to Kaluza-Klein theory.

8.3 Summary Comparison of Conventional and Topo-Phenomenological Approaches to Kaluza-Klein Theory

We know that the extra dimensions of conventional Kaluza-Klein theory are determined simply by expanding the analytic continuum in which equations for vibrating strings are written while meeting the criterion of obtaining sensible probability values. This procedure is purely formal and driven by practical considerations. Extra dimensions are added as needed, the tacit assumption being that the classical intuition of spatial continuity applies to them. Extending the dimensional canvas in this abstract way results in a rather fuzzy picture, since the general equations allow for a great many possible solutions. The details must then be arrived at indirectly, by ad hoc methods that prove awkward and unwieldy.

As in conventional superstring theory, the topo-phenomenological approach can be said to yield six additional dimensions beyond those of the known 3+1-dimensional universe. Yet the dimensions in question are not determined merely by extending space in new directions under the old intuition, and, in the process, maintaining the separation of space from the topological content therein, which content then must be specified through cumbersome ad hoc procedures. Instead space and its contents are intuited "as together" (Merleau-Ponty) thereby providing the details given in our topodimensional matrices, where there are no free parameters that require recovery.

Is it in fact appropriate to portray phenomenological Kaluza-Klein theory as "ten-dimensional"? Although its space-time dimensions do add up to ten, it is actually somewhat misleading to describe the theory in this way. For, the fundamental reality of the theory lies in the stratified pattern of sub-ten-dimensional space-time couplings, these couplings being spun out by associated pre-spatiotemporal eigenvortices. As in the standard interpretation, the 3 + 1-dimensional sphere corresponds to the visible universe, with the other dimensions being concealed from us. The presumption of the conventional view is that all ten dimensions were once on an equal footing and symmetrically related, but that, at an early point in the history of the universe, this symmetry was somehow "spontaneously broken," four dimensions expanding to visibility and the others left behind. If we do not feel entirely satisfied with this explanation of why some dimensions are hidden and others are not, looking to the mathematical theory itself for guidance will not provide us with much help. From the phenomenological standpoint, however, the selective concealment of dimensions has a distinct basis. On this account, all dimensions are *not* on an equal footing at the outset. Rather, subtle distinctions already exist between lower- and higher-dimensional wave phenomena, with the lower *carrying* the higher — a support role that eventuates in the *concealment* of the lower dimensions (see previous chapter). In the phenomenological approach then, the eclipse of the dimensions that lie beyond the 3+1-dimensional visible universe is not taken as the outcome of a "spontaneous happening" but is understood as an intrinsic feature of the theory.

It seems clear that the phenomenological version of Kaluza-Klein theory provides more detail on dimensional patterning than does the conventional rendition. In the former, the specific topological structures of the dimensions and their stratification pattern are intuitively evident. We can also say that while the extant theory affords no inherent basis for deciding which of the dimensions are spatial and which temporal (it is generally just assumed that all dimensions are spatial except for the single dimension of time that is known to us), on the phenomenological interpretation, each of the three non-zero-dimensional realms generated pre-spatiotemporally is a space-time domain unto itself, giving three distinct orders of temporality.

Naturally, since our alternative account is dialectical, its greater specificity does not simply translate as greater certitude and positivity. In the sense that our understanding of the workings of nature is enhanced by employing phenomenology, we may indeed speak of greater "positivity" — with the proviso that we grasp this positivity as inseparably blended with negativity. For the topodimensional vortices do have their *holes*, their aspects of discontinuity and indeterminateness that cannot be reduced to mere positive expression. Operating in the "forward gear," we might hope that we could mine the rich details of phenomenological Kaluza-Klein theory without sacrificing the classical aim of achieving apodictic knowledge. But reaping the benefits of the radically non-classical approach requires *shifting* gears, switching from "forward" to "backward," from projection to proprioception, from the classical intuition of objective unity to the spatio-sub-objective, phenomenological intuition of nature's *unity-in-diversity* (certainty-in-uncertainty, continuity-in-discontinuity, etc.). The price to be paid for the wealth of knowledge that is offered in the new approach is the willingness to apprehend the world in a whole new way.

Note

1. The weak nuclear force is actually carried by three distinct bosonic variations, designated W^-, W^+, and Z^0. For the purposes of this book, we will group these particles together using the collective term, "W, Z."

Chapter 9
Cosmogony, Symmetry, and
Phenomenological Intuition

9.1 Conventional View of the Evolving Cosmos

Human history can be read on many time scales. We have personal, cultural, and phylogenetic histories, and, as dwellers on this planet, we are part of the history of the earth. Moreover, being inhabitants of the universe at large, we are made up of particles that echo back to cosmic beginnings. Cosmology thus provides a way for us to explore our deepest origins. In establishing an intimate connection between the history of the universe (cosmogony) and the unity of nature, the last forty years of cosmological research have confirmed that the quest for knowledge of our ultimate source is part and parcel of a quest for harmony and accord. This, in turn, reflects the drive toward individuation we have been considering in this book.

In the currently prevailing view of cosmic history, we begin with the Planck era, a time when the entire universe was scaled below the Planck length, squeezed into the miniscule dimension of 10^{-35} meter. This was an era of quantum gravity, with the four forces of nature purportedly aligned in a relationship of perfect symmetry. This "golden age" of cosmic unity was short-lived indeed. It predates the Planck time, when our universe was but 10^{-43} seconds old. To appreciate just how young the cosmos was before the Planck time, realize that the Planck time itself is the amount of time it takes for light to travel the Planck length. When traveling at the speed of 186,000 miles per second, it does not take very long to cover a distance of 10^{-35} meter!

In conventional cosmology, the Planck time is seen to mark the initial breaking of force symmetry. The gravitational force now "freezes out." No longer blending indistinguishably with the other three forces of nature, gravitation emerges from the ultra-hot cosmic soup as a distinct force unto itself. This initiates a period that lasts from around 10^{-43} to 10^{-36} seconds after the cosmic origin, and is termed the GUT era, the stage described by physics' Grand Unified Theories wherein the three non-gravitational forces are symmetrically linked. As this epoch reaches its conclusion, a rapid inflation of the universe is said to occur. The brief expansive burst lasts roughly until 10^{-34} seconds, the universe cooling considerably during this time. The cosmos then reheats after which it begins expanding once more, now at a much slower rate. Viewed in terms of Kaluza-Klein theory, we may relate the inflation of the universe commencing at 10^{-36} seconds to *dimensional* symmetry breaking: the ten originally Planck-scaled physical dimensions diverge, four of them expanding to visibility to form the 3+1-dimensional space-time world familiar to us today.

Another important event occurs at the end of the GUT era. Force symmetry is broken for the second time. The strong nuclear force now "freezes out" of the erstwhile amalgam of non-gravitational forces, leaving only the weak force and electromagnetism as united. This marks the beginning of the electroweak era, running approximately from 10^{-36} to 10^{-12} seconds after the origin. Finally, at about 10^{-12} seconds, symmetry breaking is completed when the weak force separates from electromagnetism. All four forces presently assume distinctive forms.

Now, while the foregoing account of cosmogony touches on both force-field symmetry breaking and dimensional symmetry breaking, it is clear that these two forms of symmetry transformation are not precisely coordinated with each other in the theoretical reckoning. This reflects the fact that contemporary Kaluza-Klein theorists have been unable to articulate a detailed geometric rendering of cosmic evolution. At first speaking optimistically about Kaluza-Klein theory, the physicist Heinz Pagels noted that, "Remarkably, the local gauge symmetries are precisely the symmetries of the compact higher-dimensional space. Because of this mathematical fact, all the gauge theories of Yang-Mills fields can be interpreted purely geometrically in terms of such compact higher-dimensional spaces" (1985, pp. 327–28). However, for the geometric program fully to be realized, the physical events described in the standard and inflationary models of cosmic development would need to be specifically expressible as dimensional events. "Unfortunately," admits Pagels, "no one has yet been able to find a realistic Kaluza-Klein theory which yields the standard model" (p. 328). We have already discussed the reason why extant Kaluza-Klein theory has not been able to fulfill its geometric program by specifying the exact shapes of the hidden dimensions that would correspond to the physical facts of the standard model (see previous chapter). Of course, if the prevailing theory cannot tell us what the dimensional structures are that correspond to physical reality, it can hardly inform us on how these dimensions develop. In point of fact, there is really no positive feature intrinsic to the theory that provides for the evolution of dimensions. As far as I can tell, the only reason dimensional bifurcation is assumed to have taken place at all is that theorists must somehow account for the present inability to observe six of the ten dimensions needed for a consistent rendering of quantum gravity (one that avoids untenable probability values).

9.2 The Problem of Symmetry

The crux of the problem with established cosmological theory can be viewed in terms of its core assumption about symmetry. As I noted in Chapter 6, conventional Kaluza-Klein theory puts symmetry first, taking it as primordial. Cosmogony is described as a process in which an original "perfect symmetry" is broken. What I demonstrated in the earlier chapter is that the primacy given to symmetry reflects an underlying intuition of changelessness that is essentially incompatible with any notion of genuine cosmic change.

The writings of the physicists Heinz Pagels and Brian Greene well exemplify the way symmetry is privileged in contemporary physics. Pagels's (1985) title alone speaks eloquently to this bias: *Perfect Symmetry: The Search for the Beginning of Time*. For his part, Greene notes that, "Over the last few decades, physicists have elevated symmetry principles to the highest rung on the explanatory ladder" (2004, p. 224). "Symmetry underlies the laws of the universe," says Greene, and "*everything* we've ever encountered is a tangible remnant of an earlier, more symmetric cosmic epoch" (pp. 219–20). Continuing in the same vein, he asserts that "the universe we currently see exhibits but a remnant of the early universe's resplendent symmetry" (p. 268). But just how accurate is this picture of an originally symmetric universe, one that then evolved through symmetry breaking?

It is not only *contemporary* physics that privileges symmetry. Physics has always done so. What we found in the opening chapters of this book is that physicists, from the outset, have been looking to subdue the inhomogeneous, chaotic element of nature as part of their ongoing quest for individuation. We considered this basic nonsymmetric feature under its ancient name, the *apeiron*. Plato wrestled with the *apeiron*, seeking through the action of his Demiurge, and through his incipient notion of space (the "receptacle"), to bring order to an initially "disorganized state" in which "there was no homogeneity or balance...[no] equilibrium...[no] proportion or measure" (1965, p. 72) — no symmetry, that is. Armed with a more advanced concept of the spatial continuum, Descartes and Newton did better than Plato in containing *apeiron*. In Newtonian physics, nature's concrete divergence from ideal symmetry was small enough to be effectively ignored. Then, with the late nineteenth century findings of Michelson and Morley, and of Planck, the old symmetries were swept away necessitating the introduction of new, more abstract symmetries, those of relativity and quantum mechanics. In all this work, the underlying aim is to contain the nonsymmetry of apeironic nature, to impose symmetry upon it (see Rosen 2004).

We have considered the most ambitious effort in this regard: superstring theory. In cosmological terms, the further back we go into the early universe, the greater the turbulence we encounter. String theory aspires to deal with the embryonic state of affairs prevailing around the Planck time, when the universe was only 10^{-43} seconds old. Where the standard model of cosmology cannot cope with the tumultuous nonsymmetries of this period, string theory hopes to succeed in "calming" these "violent quantum jitters" (Greene 1999, p. 158). The effectiveness of conventional string theory is questionable, as we have seen. But even if the symmetries of the theory could be worked out more definitively, would this mean that the universe itself was highly symmetric around the Planck time, as Greene seems to suggest? Or would it be more accurate to say that theorists had achieved an abstract symmetric account of a situation that, in concrete reality, was highly nonsymmetric?

Indeed, the equations of string theory, and of contemporary physics in general, are expressly designed to isolate mathematically whatever actual nonsymmetries are encountered, to cordon them off in the interest of realizing abstract symmetry (see Chapter 4). Modern physics primarily has

been concerned not so much with describing the inhomogeneities of chaotic nature as they actually are, but with "quarantining" those nonsymmetries, relegating them to a "black box" so that nature can be described via
theoretical symmetry. Perhaps we should say then that the symmetries of
cosmological theory do not reflect the concrete facts of the restless early
universe as much as the effort to subdue that restlessness via present-day
mathematical abstraction. If that is true, would it not be an error of mis-
placed concreteness to attribute "perfect symmetry" to the early universe?
At least with the pragmatic positivism of the Copenhagen Interpretation,
there is the honest admission that, while the rarified symmetries of quantum mechanics might work to a high order of precision in predicting laboratory outcomes, they have nothing meaningful to tell us about underlying
reality. But contemporary cosmology's allegedly "realistic" interpretation
clouds the issue by creating the impression that its theoretical equations
are describing the actual situation in the high-energy nascent universe,
when in truth it is doubtful that they are.

In fact we can agree that, at "high enough temperatures…there is no
distinction between the weak nuclear force and the electromagnetic force"
(Greene 2004, p. 265) — or between any of the forces of nature, for that
matter. I venture to say, however, that the primordial indistinguishability
of the forces actually does not reflect a symmetric relationship among different fields, but a pre-symmetric state of affairs in which there are neither
well-defined differences nor stable identities.

The alleged symmetry of the early universe has been likened to the
phenomenon of ferromagnetism. If a magnet is heated, the many small
magnets of which it is composed become agitated and begin jiggling about
erratically. When this happens, the mini-magnets lose their common orientation and their axes become randomly distributed. With no preferred
orientation, the many different orientations can be described as "canceling
each other out." It is this overall absence of difference that is said to constitute symmetry. Yet, while statistical generalization may be mathematically convenient, the concrete reality of the individual mini-magnets is
that of continual variability. At least at this level we hardly have symmetry, if symmetry is defined in terms of *in*variance.

But let us look more closely at the probabilistic approach. The tossing
of a coin gives the simplest illustration of the principles involved. Here

there are only two possible outcomes: heads or tails. If the coin is evenly balanced and is randomly flipped, there is no way of knowing in advance what the outcome will be. However often the coin is thrown, there is the same uncertainty about the outcome of any particular event prior to its actual occurrence. Naturally, there is never any doubt about whether a coin that has already been tossed *is* a head or a tail; the uncertainty only enters prior to the toss, when we cannot know beforehand what the result will be. Probabilistic analysis thus presupposes complete certainty with respect to the value of any particular actually occurring event. The uncertainty does not have to do with the given event per se but with which of two or more events will occur when the possibilities are distributed such that no particular event is determinately favored, each event having a positive likelihood of occurring.

If we adopt a more general perspective, focusing not on the individual events themselves but on the overall distribution of events, we find that, in the coin-tossing example, with increased numbers of tosses the numbers of heads and tails converge ever more closely. Thus differences cancel each other out and we approach statistical equilibrium or symmetry. In sum, probabilistic analysis takes it for granted that each event has an unambiguous value when it actually occurs, and it assumes that — while there is ambiguity over which event will occur on any given occasion before the event is actualized, when the results of many such concrete occasions are considered from an abstract statistical standpoint, uncertainty approaches certainty, probabilistic invariance.

This is the basic symmetry idea that is conventionally applied to the "quantum soup" (Pagels 1985, p. 252) that constitutes the early universe. Like the differently oriented coins or the differently directed components of the heated magnet, the particles immersed in the hot quantum soup are taken as unambiguously valued events in themselves that are randomly distributed with respect to each other, these chance fluctuations being symmetrically interchangeable in relation to the overall statistical equilibrium that prevails. It is in this sense that the early universe is deemed "perfectly symmetric."

From the depth-dimensional standpoint, of course, the most primordial state of affairs is not one of self-symmetric identity, but of apeironic chaos. In this book and elsewhere (Rosen 1994, 2004), I have proposed that the probabilistic approach of quantum mechanics is a stratagem designed to mask the inherently discontinuous, indeterminate, nonsymmetric aspect of nature present from the origin. We can now see exactly how this is done. The classical ideal of the atom — the self-identical, internally invariant individual — is upheld despite frenzied microworld variability by externalizing the uncertainty residing within the concrete core of the individual event. Core uncertainty is transformed into an uncertainty merely concerning which among several randomly distributed particle events will occur, with each such event itself assumed to be internally stable and determinate. Once uncertainty is abstracted in this way, it can then be resolved by the law of large numbers: though the occurrence of a particular event cannot be determined with complete certainty in advance, certainty (equilibrium, symmetry) is asymptotically reached with regard to the overall distribution of events. In Chapter 2, we saw how essentially the same method is used to uphold the notion of spatial locality on which the ideal of the continuum depends: the intrinsic non-locality of the particle is denied by transforming the core uncertainty about the particle's spatial position into an external uncertainty concerning which among several possible positions the particle is located at, with each location itself being assumed determinate and overall probabilistic closure being attained.

But we have also discovered that probabilistic closure becomes progressively more difficult to maintain as we approach more closely the turbulent energies associated with the Planck length. The probabilistic masking of inherent nonsymmetry in unified field theory has had its greatest success at the relatively low energies of electroweak unification. Cosmogonically, this corresponds to the state of affairs existing before the universe was 10^{-12} seconds old, the time when the weak force "froze out" from its union with electromagnetism. The "quantum soup" of photons and weak gauge bosons was surely hot in the electroweak era (10^{15} K) but was not "boiling over." The fluctuating particle events therefore could still be considered as internally stable, finite deviations from statistical equilibrium. What I am proposing is that the mathematically conceived fusion of force particles was in actuality a *con*fusion, that — instead of a symmetrical exchange of particles with internally well-defined values, there was a pre-(a)symmetric flux in which particles had no definite internal values, in which there were no simply self-identical particles, no "ultimate atoms," no fixed individuality. Here we can rightly speak neither of particles with different properties, nor of particles with the same properties. It is this pre-(a)symmetric state, where we have neither asymmetry nor symmetry, that the probabilistic analysis seeks to contain by rendering it "symmetric." And electroweak theory did achieve a measure of success on this score.

However, as we draw closer to the Planck time of 10^{-43} seconds and the quantum soup becomes hotter, individual fluctuations are character-ized more and more by violent agitation, so that suppression of concrete indeterminacy, the core indeterminacy of the single actual event, is harder to enforce. Grand Unification Theory would impose symmetry on the $>10^{28}$ K roiling soup of gluons, photons, and weak gauge bosons existing earlier than 10^{-36} seconds after the origin of the cosmos. This attempt at unification has indeed run into greater problems than electroweak unifica-tion and has still not been brought to successful completion. Of course, the greatest theoretical difficulties have been encountered for the $>10^{32}$ K quantum gravitational soup existing before 10^{-43} seconds, and we have discussed the problems of superstring theory in this regard. String theory confronts a situation where event fluctuations are so great that statistical closure is virtually unattainable under the assumptions of the standard model and probability values accordingly prove untenable there. This is where divergences cannot be made to "settle down"; instead of canceling each other, averaging out toward statistical equilibrium, they grow to the point where the inchoate flux of *apeiron* cannot be masked. It is true that string theory, for its part, can manage to achieve probabilistic closure by writing its equations in ten dimensions. We have found, however, that the mathematical uncertainty avoided in the general equations reappears as an epistemic uncertainty in solving those equations, since the equations per-mit a great many possible solutions and provide no intrinsic means of re-ducing the field of possibilities so as to align theory with physical reality.

Despite these questions about the applicability of the symmetry idea to the very early universe, the depth of science's commitment to this con-cept cannot be overstated. The way Greene puts it,

> Physicists...believe [their] theories are on the right track because, in some hard-to-describe way, they *feel* right, and ideas of symmetry are essential to this feeling. It feels right that no location in the universe

is somehow special compared with any other....It feels right that no particular...motion is somehow special compared with any other....So the symmetries of nature are not merely consequences of nature's laws. From our modern perspective, symmetries are the foundation from which laws spring. (2004, p. 225)

We can say then that, for contemporary physics, symmetries are much more than patterns observed in nature. The symmetry principle constitutes the *intuitive basis* for all theoretical formulations of nature. According to this root intuition, "it feels right" that something abstract should stay the same amid concrete changes. This sense of "rightness" is surely well-embedded in the mindset of Western science and philosophy, dating back to the Platonic forms, the abstract universals that remain invariant despite countless variations in their particular exemplars. Unmistakable here is the old intuition of object-in-space-before-subject. The abstract universal or "eternal object" is the transcendent subject's invariant model or archetype, the changeless symmetry that mediates between the subject per se and the transitory concrete objects that are contained in space. It is the Platonic predilection of contemporary mathematical physics that makes it "feel so right" to privilege changeless form over concrete change. More basically, the preeminence of the symmetry principle reflects the drive toward unity when the process of Individuation is proceeding in the "forward gear." But while physicists may thus feel strongly compelled to put abstract symmetry first, what I am claiming is that, if they continue to do so, an understanding of our cosmic origins will continue to elude them. Before further specifying my proposed alternative, let me explore in greater depth its radical departure from the standard approach.

9.3 A New Kind of Clarity

The problems of cosmology might not be so difficult to solve if all that we required were the right theory, as modern physics seems to assume. Operating under this assumption, many physicists feel optimistic that a "Theory of Everything" could be just around the corner. However, the upshot of my presentation is that — more than a new theory, a fundamentally new *intuition* is needed for an effective approach to the questions of cosmic evolution and cosmic unity (see Rosen 1994, Chapter 5). It should be clear then, if I am right, that the change in outlook will have to go deeper than is currently suspected in contemporary physics. A whole new philosophical foundation will be necessary. And I am suggesting that this foundation be *phenomenological*.

It is easy to underestimate the true magnitude of the change that is in order. In the closing section of the previous chapter, for example, I indicated that the phenomenological version of Kaluza-Klein theory clarifies the conventional rendition by providing more detail on dimensional patterning. But it would actually be more accurate to say that phenomenology offers a *new kind of clarity*, one in which the idea of "detail" has a rather different meaning.

Assuming the conventional posture of mathematical physics, we begin from an intuition of abstract universality, viz. the symmetric forms that constitute the general equations. We then look to solve those equations so as to determine the particular structures that can be put into correspondence with physical reality. However, the general and particular levels of the theory are separated from each other in such a way that the sense of intuitive confidence felt about the abstract symmetries does not carry over into the detailed solutions of those equations. That is why, in string theory, we are unable to say which of the many possible topological solutions "feels right" to us. Physicists have had to play the elaborate guessing games we have discussed because their mathematical formulations afford no intuitive guidance as to the right shape of the hidden dimensions, and because those dimensions defy direct empirical observation. The "devil" surely has been in the details.

With phenomenological intuition it is different. Here the generalities intuited are inseparable from the details, since these general structures are not abstractions that must be concretized in a subsequent step but are "general *things*," universals that themselves are concrete. So, when I say that the phenomenological version of Kaluza-Klein theory gives more detail on dimensional patterning than the conventional version, I am not speaking of "details" in the usual sense of particular features or instances of a pattern that itself is more general, the *division* of the particular and general being tacitly assumed. Rather, I am referring to a pattern whose details are an integral part of its very generality.

This certainly makes for a different kind of clarity than is customarily sought. Conventional clarification can be further understood in terms of figure-ground relations, where it can be likened to the process of bringing into focus a particular image from a general background that recedes from awareness when the figure emerges. I glance around my room, diffusely cognizant of all the objects it contains. Then, from this global impression, I bring my attention to one object in particular, the picture on the far wall, whereupon the objects in the periphery are largely excluded from my awareness. Could I counteract the mutual exclusion of figure and ground and apprehend the background *at the same time* that I focus on a given object? Normally this is deemed impossible. Yet experimentation with a certain curious figure from the psychology of perception suggests there can be another answer.

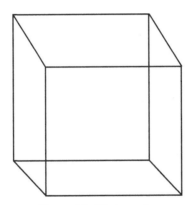

Figure 9.1. Necker cube

The Necker cube (Fig. 9.1) is a reversible figure. You may be perceiving it from the point of view in which it seems to be hovering above your line of vision when suddenly a spontaneous shift occurs and you see it as if it lay below. Two distinct perspectives thus are experienced in the course of gazing at the cube, yet the cube's reversing viewpoints overlap one another in space, are internally related, completely interdependent (think of what would happen to one perspective if the other were erased). It is because the objects in my room appear to be simply separated from each other that, in focusing on one, I am largely oblivious to the background that encompasses the others. In the case of the cube, however, the more intimate relationship between perspectives means that when I gaze at one of them, the backgrounded perspective is in close proximity; therefore, in viewing the figure, the ground is less opaque to me than in ordinary viewing. As I focus on one Necker cube perspective, the peripheral perspective "translucently" shines forth. Can the background become fully *transparent* as I view the figure?

Glimpsing opposing perspectives of the cube "transparently" would mean that, rather than being limited to one static viewpoint of the cube or the other, we would directly apprehend what lies dynamically *between* them. Let us reconsider the actual manner in which we normally become aware of the cube's reversing perspectives. We are viewing the cube from one point of view when, quite suddenly, we realize that we are now perceiving it from the other. When this happens, we are not actually cognizant of the transition itself, but only of the result of this event. That is, in deducing retrospectively that a shift "must have taken place," we engage in an abstractive reconstruction of the concrete act, rather than directly apprehending it. Is there another manner of experiencing the cube? Might there not be a way to depart from perceptual habit and gain an immediate insight into the actual reversing of opposed perspectives?

Merleau-Ponty categorically denied this possibility. For him, the "hinge" between reversible perspectives is opaque; it is "solid, unshakeable ...irremediably hidden from me" (1968, p. 148). Many students of perception would agree. For example, Asghar Iran-Nejad (1989), after reporting on psychological research said to unequivocally demonstrate the mutually exclusive fashion in which the cube's two perspectives are perceived, buttressed his empirical finding with a logical argument: "To the extent that two different schemas must share [the same] knowledge components, they cannot...be held in mind simultaneously just as two meetings cannot be held at the same time to the extent that the same individuals must participate in both of them" (1989, p. 131). This point of logic boils down to the *principle of non-contradiction*, characterized by philosopher Peter Angeles as one of the basic "laws of thought" in Aristotelian philosophy: "A thing *A* cannot be both *A* and not *A* (at the time that it is *A*)" (1981, p. 153). Angeles further observed that classical logic has been challenged by *dialectical* logic, "a logic of becoming that attempts to present the ever-changing processes of things" (p. 155). Here "contradiction exists in reality. It is possible for the selfsame thing to be and not to be — the same thing is and is not" (p. 155). Since there would be no contradiction in saying that *A* becomes not-*A* with the passage of time, dialectically it is understood that *A* is not-*A* "at the time that it is *A*" (p. 153).

The dialectical lifeworld certainly does not conform to Aristotelian strictures. We know from Chapter 3 that lifeworld temporality is not the mere sequence of point-like, extensionless "nows" posited by Aristotle. It is only if time were Aristotelian that *A* and not-*A* could not be identical to one another in the same instant. But the "true time" adumbrated by Heidegger possesses the dialectical thickness in which the "future, past, and present...belong together in the way they offer themselves to one another" (Heidegger 1962/1972, p. 14). Thus, in the temporality of the lifeworld, it is indeed possible for opposing perspectives of the cube to coincide "simultaneously" (see below for a significant qualification of this term). And it is possible for us to *experience* this concurrence, assuming we do not impose an artificial restriction on our experience. Still, we do have trouble recognizing these possibilities. This is not surprising given the depths of our roots in the classical ground. Understandably, our perception of the Necker cube is strongly influenced by the classical posture that has long governed human functioning. Therefore, even though we are able to reconstruct abstractly the concrete process of reversing perspectives, we find it difficult to grasp that process directly — difficult, but not impossible.

Instead of allowing our glance to alternate between the opposing perspectives of the cube, we can break this visual habit and view both perspectives at the "same time." I have explored this possibility elsewhere (Rosen 1988, 1994, 1997, 2004), and it is confirmed in Bruno Ernst's (1986) study of *Belvedere*, a graphic work by M. C. Escher (Fig. 9.2).

In his analysis of the Escher print, Ernst calls our attention to a detail at the bottom (a little left of center): a boy is sitting on a bench and puzzling over an odd-looking structure he holds in his hands. Lying at his feet is the blueprint for this structure, and, indeed, for the entire artwork. It is the Necker cube. To bring out the underlying principle of *Belvedere*, Ernst provides his own diagram of the cube (Fig. 9.3).

Figure 9.2. M.C. Escher's *Belvedere*

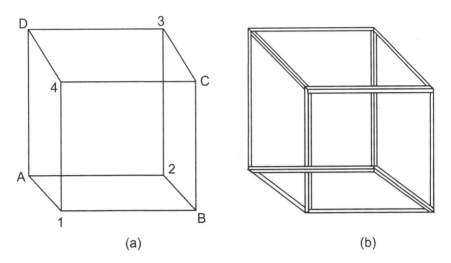

Figure 9.3. Bare Necker cube (a) and cube with volume (b) (after Ernst 1986, p. 86)

According to Ernst, the cube encompasses within itself,

> the projection of two different realities. We obtain the first when we assume that points 1 and 4 are close to us and points 2 and 3 are further away; in the other reality, points 2 and 3 are close and 1 and 4 further away....But it is also possible to see points 2 and 4 in the front and 1 and 3 in the back. However, this contradicts our expectation of a cube; for this reason, we do not readily arrive at such an interpretation. Nevertheless, if we give some volume to the skeletal outline of the cube, we can impose said interpretation on the viewer by placing A2 in front of 1–4 and C4 in front of 3–2. Thus we obtain [Figure 9.3(b)] and this is the basis for *Belvedere*. (Ernst 1986, p. 86; translated by M. A. Schiwy)

It is clear that Figure 9.3(b) "contradicts our expectation of a cube" because it brings together opposing "realities" (perspectives) that we are accustomed to experiencing just one at a time. When this happens, there is an uncanny sense of self-penetration; the cube appears to do the impossible, to go *through* itself (thus Ernst speaks of constructions based on the cube as "impossible"; 1986, pp. 86–87). If we again imagine the bare cube (Figs. 9.1 and 9.3(a)) as a solid object appearing in space, one whose faces are filled in, we find that perspectival integration has an interesting effect on those faces.

In the conventional, perspectivally polarized way of viewing the cube, when the shift is made from one pole to the other, all the faces of the cube that had been seen to lie "inside" presently appear on the "outside," and vice versa. But it is only at "polar extremes" that faces are perceived as *either* inside *or* outside. With the fusion of perspectives that discloses what lies *between* the poles (the "between" that Merleau-Ponty judged "irremediably hidden"), each face presents itself as being inside and outside at the same time. Therefore, the division of inside and out is surmounted (at least symbolically) in the creation of a *one*-sided experiential structure whose opposing perspectives are simultaneously given.

Simultaneously? Well, that is not exactly the case. I am proposing that we can apprehend the cube in such a way that its differing viewpoints overlap in time as well as in space. But what we actually experience when this happens is not simultaneity in the ordinary sense of static juxtaposition. We do not encounter opposing perspectives with the same immediacy as figures appearing side by side in Cartesian space, figures that coexist in an instant of time simply common to them (as, for example, the letters of the words on this page). If the coincidence of the cube's perspectives were limited to that, Merleau-Ponty certainly would be right to say that such opposites could "never reach coincidence" (1968, p. 147). But there is indeed a coincidence in the integrative way of viewing the cube, for perspectives are not related in simple Aristotelian succession (first one, then the other) any more than in Cartesian simultaneity. If opposing faces are not immediately co-present, neither do they disclose themselves merely *seriatim*, in the externally mediated fashion of linear sequence. Instead the relation is one of internal mediation, of the *mutual permeation* of opposites. Perspectives are grasped as flowing through each other in a manner that blends space and time so completely that they are no longer recognizable in their familiar, strictly dichotomized forms. You can see this most readily in viewing Figure 9.3(b). When you pick up on the odd sense of self-penetration of this "impossible" figure, you experience its two modalities neither simply at once, nor one simply followed by the other, as in the ordinary, temporally broken manner of perception; rather, you apprehend an unbroken flowing from one to the other.[1]

Note that to view figures in simple simultaneity is to view them as merged into a background to some degree, so that a measure of detail is lost. Viewing the objects in my room in the globally juxtaposed manner described above, the sharp definition of the individual objects is sacrificed; viewing the words of this paragraph all at once, individual words blur together; and viewing the letters of a single word as a simple whole, individual letters lose resolution. But with the perspectival integration of the Necker cube, the non-simultaneous aspect assures that focal clarity is maintained. Distinct sides of the cube do not just dissolve into a global background, for, in the experience of integration, faces of the cube *are* inside, and yet outside as well. Without concrete detail simply vanishing, distinctive perspectives flow through each other in mutual transparency to bring a unity or universality that encompasses the background in its sweep. In this sense then, figure and ground are perceived as together; the focal and the peripheral, the concretely particular and the universal, are encountered in dynamic harmony.

However, we must now consider an obvious limitation in the exercise with the Necker cube. In the ordinary three-dimensional context in which our viewing takes place, the cube is but a one-dimensional line drawing embedded in the two-dimensional space of the printed page. Therefore, while the perception of the cube does simulate our experience with solid entities in three-dimensional space, the cube itself is certainly no solid. I am proposing that the perspectival integration of the cube conveys a sense of the lifeworld's concrete universality. But it is clear that such an insight falls short of encompassing the actual lifeworld of three-dimensional objects and events. Therefore, although the non-Aristotelian apprehension of the cube surely may *symbolize* the dialectical transpermeation of inside and outside — and of subject and object, it is *no more than* symbolic. But by now we are well aware of a one-sided dialectical form sufficiently dimensioned to embody the actuality of the three-dimensional concrete universal. I am speaking of the phenomenologically intuited Klein bottle. It is this structure that gives adequate expression to the new kind of intuitive clarity necessary for an understanding of the physics and cosmology of our three-dimensional world. Of course, a full understanding of cosmic evolution requires that we take into account the lower-dimensional concrete universals corresponding to the lower-order members of the topological bisection series. It turns out that the Necker cube too is a member of a dialectical series, and that the several orders of this visual dialectic parallel the topological orders we have been considering. I have examined these parallels elsewhere (Rosen 2004, pp. 183–84) but will refrain from exploring them here in the interest of keeping the scope of this book within manageable limits. My only reason for introducing the Necker cube in the present section is to convey to the reader in the simplest way an intuitive sense of how phenomenological clarification at once encompasses both figure and ground, part and whole, concrete detail and global context. Having done that, I am now prepared to elaborate my proposed phenomenological alternative for the story of cosmic creation.

Note

1. Although Merleau-Ponty did not seem to be in contact with this dialectic of mediation and immediacy in the material I have cited above, he appeared to approach it elsewhere. See his discussion of "self-mediation" (*médiation par soi*) (1968, p. 92).

Chapter 10
The Self-Evolving Cosmos

10.1 Introduction to the Cosmogonic Matrix

Cosmological theory would take a giant step forward if its geometric and physical descriptions of the evolving universe could be brought into correspondence. But this step is blocked by the primacy given to the concept of symmetry, since, with symmetry coming first, intrinsic dimensional change is essentially precluded. In the alternative I am proposing, our understanding of dimensional change is not rooted in the notion of symmetry breaking that puts symmetry first, but in the dynamic intuition of symmetry *making* (see Chapter 6).

Dimensional generation basically entails the development of a dimension's eigenstructure via distillation of its boundary or identity elements. Of course, for *lifeworld* dimensionality, these dimensional features cannot be lifeless geometric abstractions but must be individual events, "actual occasions," to use philosopher Alfred North Whitehead's (1978) term.[1] In physics, the individual events are the fundamental particles and fields, which gain dimensional expression in a Kaluza-Klein context. It is by assuming the dimensional particle event to be *simply* individual — simply indivisible or atomic from the outset; self-symmetric, thus inherently static; closed into itself and therefore impervious to change at its core — that an intrinsically geometric account of the evolving universe is obstructed. I have proposed that the dimensional development that underlies cosmogony concerns the process of Individuation; it is about the *forming* of the dimensional individual (the atomic particle event). This process can hardly be grasped if we begin with the assumption of a pre-existent, changeless individual.

The topo-phenomenological rendition of cosmogony takes its cue from the stages of dimensional generation set forth in Chapters 6 and 7. The matrices of Table 7.1 encapsulate these stages. In Table 10.1, we recast the earlier table so as to spell out the basic dimensional particle events and their co-evolution. Notice that the new table is formatted after the design of Table 7.1(a) rather than 7.1(b). Whereas the latter parses the circulations of the dimensional spiral, teases them apart for easier identification, the former more effectively brings out the nonlinear interwovenness of the windings. It is the intertwinement of the dimensional circulations that we must now come to better appreciate.

Table 10.1 expresses the topodimensional spinors of Table 7.1 in terms of their corresponding particle events. The several gauge bosons are found in their developed forms on the principal diagonals of the matrices: G, the graviton, corresponds to ε_{D0} of Table 7.1; g, the gluon, is linked to ε_{D1}; W and Z, the weak gauge bosons, are associated with ε_{D2}; and γ, the photon, with ε_{D3}. From the phenomenological standpoint, these are not merely finite particular entities to be objectively measured in the three-dimensional framework of the laboratory. Rather, each bosonic particle (or force field) is seen as a spatio-sub-objective dimension unto itself; each is a concrete universal or archetypal atom of process[2]; each is the action of a lifeworld or braneworld.

We can regard the several bosons as "musical" actions, vibrating dimensional "strings." The bosons constitute "fundamental tones" (Chapter 5) taking the forms of our topodimensional vortex waves. In Chapter 8, I spoke of the waves as *eigen*vortices, since the action in question is self-referential. The eigenvortical branes or "twistors" (cf. Penrose) are the prereflective, pre-spatiotemporal actions that — in twisting back upon themselves, first make manifest the space-time force fields (electromagnetic, weak nuclear, etc.) in which reflective action is possible and symmetries can be established. Yet the archetypal branes themselves are neither symmetric nor asymmetric alone, but *syn*symmetric, entailing as they do the dialectical blending of symmetry and asymmetry (whole and hole, object and subject, continuity and discontinuity, etc.). Most essentially, we may view the evolution of the cyclonic dimensional waves as the working out of this dialectic.

10.2 Overview of Cosmic Evolution

Before examining stage by stage the details of cosmogony that are given in Table 10.1, let us consider the overall pattern of dimensional genera-

Clockwise Stages (Projection):

Stage 4
(γ)
Present Era
$t > 10^{-12}$ sec.

Stage 3
$(W, Z; \gamma)$
Electroweak Era
10^{-36} to 10^{-12} sec.

Stage 2
$(g; W, Z; \gamma)$
GUT Era
10^{-43} to 10^{-36} sec.

Stage 1
$(g; W, Z; \gamma)$
Planck Era
$t < 10^{-43}$ sec.

Counterclockwise Stages (Proprioception):

Stage 5
(γ)

Stage 6
(γ)

Stage 4
(W, Z)

Stage 7
(γ)

Stage 5
(W, Z)

Stage 3
(g)

Stage 8
(γ)

Stage 6
(W, Z)

Stage 4
(g)

Central grid:

			γ $[e, \nu_e]$
		W, Z $[\mu, \nu_\mu]$	$(W, Z)/\gamma$ $[u, d]$
		$\gamma/(W, Z)$	
	g $[t, b]$	$g/(W, Z)$ $[\tau, \nu_\tau]$	g/γ $[c, s]$
	$(W, Z)/g$		
	γ/g		
G	G/g	$G/(W, Z)$	G/γ
g/G			
$(W, Z)/G$			
γ/G			

Table 10.1. Stages of cosmogony

tion. As with Table 7.1(a), Table 10.1 displays the full course of development of all orders of topodimensional action. What I said in Chapter 7 about Table 7.1(a) applies here: The general design of the Table is circular. Moving upward from the matrix at the bottom, we have the clockwise, forward, or projective stages of braneworld generation, indexed by the terms appearing to the left of the matrices. The turn to the counterclockwise, backward, or proprioceptive stages is then enacted by moving back down through those same matrices, with stage numbers now displayed to the right. The parenthetic terms accompanying each stage number indicate the bosonic braneworlds to which that number applies; since the zero-dimensional graviton does not undergo development, it does not appear here. Note that, if the sequence of stages for each braneworld is considered separately from that of the other braneworlds, Table 10.1 in fact does not describe the action of a single circle but of circles nested within circles, so that the overall pattern is actually that of a *spiral*.

Now, owing to the linearity of Table 10.1, the four matrices are shown as simply separated from one another. The impression is thereby conveyed that, in advancing from one matrix to another, the matrices that are passed are merely left behind. Thus, in the projective stages, the graviton appears simply to drop out in the transition to stage 2, the gluon seems to disappear in stage 3, and the weak gauge bosons to vanish in stage 4. Instead of reading the progression of matrices in this linear fashion, let us read the matrices as *overlapping* one another. Such a reading seems warranted by the fact that actual stage-to-stage movement cannot be linear in nature but must be paradoxically discontinuous. Yes, the divisibility of the linear space-time continuum is *generated* in the course of dimensional development, but by a form of action that is itself indivisible. Therefore, rather than involving a continuous passage between juxtaposed points that are externally related to one another, cosmogony entails a "quantum leaping" to new modes of integral action, with each novel mode related to its predecessor in an *internal* way.

The overlapping entailed in discontinuous phase transitions is well modeled by the visual structure employed in the previous chapter: the Necker cube (Fig. 9.1). In taking the leap from one perspective of the cube to the other, the first perspective is eclipsed (assuming we are not dealing with the full *integration* of perspectives; see Chapter 9 and below). And yet, because of the intimate interdependence of perspectives, the occlusion is incomplete. As Merleau-Ponty put it, "the hidden face of the cube radiates forth somewhere as well as does the face I have under my eyes, and coexists with it" (1968, p. 140). For Merleau-Ponty, the "coexistence" of the hidden face with the visible face means an intertwinement so thorough that the "hidden" face could not *merely* be hidden; it too must fully be present, albeit in a less explicit, background fashion. This is the sense in which antecedent matrices and their braneworlds are not simply left behind with the passage from one stage of projection to another. A visual metaphor employed by Jean-Paul Sartre may help to further clarify the nonlinear transition in question. We might say that the stage-to-stage matrix reduction given in Table 10.1 signifies a stratification of functioning in which "translucent layers of consciousness" are created.

Sartre's idea of the "translucency of consciousness" (1943/1956) constitutes a rejection of the Freudian dualism of unconscious and conscious wherein the former is opaque to the latter. For Sartre, consciousness is not categorically compartmentalized in this way. He would have us think instead that, while there are areas of our awareness that are not fully transparent to us, some "light" always gets through, as if through a translucent screen. Similarly, in advancing by "quantum leaps" to subsequent stages of clockwise cosmogony, antecedent dimensional matrices are overshadowed but by no means entirely eclipsed; some "light" from each one of them manages to get through. Functioning becomes gradated in such a way that a differential penumbra is formed, with the more primordial braneworlds being shaded more darkly, yet never completely obscured.

It is no accident that the stage-to-stage matrix reduction of Table 10.1 needs to be read in this Sartrean/Necker-cube-like fashion. In concluding the previous chapter, I intimated the close relationship between the cube and the topological entities that make up the principal windings of the cosmogonic spiral. So the Necker-cube reading of the Table is a *topological* reading. Just as the quantum leap to a whole new perspective of the cube also implicitly maintains the initial perspective, the transition from one side of the Moebius strip or Klein bottle to the other paradoxically keeps us on the same side. Applying this to relations among the braneworlds of Table 10.1, we may say that, while projective phase transitions reduce the matrices by seeming to eliminate certain branes, all braneworlds actually continue to operate, albeit in a penumbral fashion in which their "light" is attenuated in direct proportion to their primordiality.

If the stage transitions of cosmogonic projection are modeled by the quantum leap from one perspective of the Necker cube to the other in which the initial perspective persists as a kind of penumbra, the stage transitions of proprioception can be related to the *perspectival integration* of the cube. Although the Necker cube is ordinarily viewed in such a way that, at any given moment, a single perspective is dominant, the other being relegated to the background, with the help of Figure 9.3(b) we saw that it is possible to go beyond the customary manner of perceiving the cube and apprehend its perspectives as dynamically integrated. The dialectical interpenetration of perspectives that results from counteracting the figure-ground tendency to project one perspective to the fore and occlude the other symbolizes the shifting of gears to counterclockwise, to the retrograde act of proprioception wherein the implicit perspective is made explicit (wherein "consciousness...[becomes] aware of its own implicate activity"; Bohm 1994, p. 232). Accordingly, if clockwise phase transitions are understood as Necker-cube-like passages that relegate previous stages to a background penumbra, we may regard counterclockwise transitions as proprioceptively reversing that forward thrust, and, in the process, bringing about an integration in which all stages, matrices, and cosmodimensional spheres become diaphanous to one another. Evidently then, whereas the clockwise matrices are to be read as superimposed upon one another in translucent layers, we are to read the counterclockwise matrices as *transparent* — not in the ordinary sense of static simultaneous visibility, but in the topo-dialectical sense of dynamic interpenetration.

Generally speaking, Table 10.1 provides a developmental account of the sixteen fundamental particles that form the basis of the objective universe according to contemporary physics. Here we are able to see how the twelve kinds of fermions or matter particles of physics' standard model (the particles signified by square-bracketed terms) are distilled from interactions among the four types of bosons or force particles (all transformations are examined in detail in Sections 10.4 and 10.5). This comprehensive rendering of cosmic evolution derives entirely from first principles, so there is no need to add free parameters to "make things work." In the matrices of Table 10.1, physical and geometric descriptions of the evolving universe are brought into full correspondence, since every particle in the Table has its specific dimensional counterpart (fermions are interpreted as dimensional bounding elements; see next section). Given that all dimensional particles originate in a clear-cut manner from the same closely-knit family of topological spinors, the matrices of Table 10.1 bring the Kaluza-Klein unification program to fruition. By the same token, string/ M-theory gains fulfillment here because the dimensional particles lend themselves to being interpreted as strings, or as their generalizations, *n*-dimensional branes. So Table 10.1 sketches the "musical score" for the symphony of dimensional spheres.[3] (For the preliminary overview provided in the present section, it will not be necessary for the reader to take in and grasp every detail of the Table. Be assured that all the terms and features of the Table will be accounted for in due course.)

In certain general respects, the clockwise stages of Table 10.1 are consistent with the picture of cosmogony given in the standard and inflationary models, and in Kaluza-Klein theory. As in contemporary cosmological speculation, we are saying that the universe began with an inestimably hot and dense "singularity" that ushered in the Planck era, a period during which all the forces of nature — fully understood here in dimensional terms, were inseparably blended. Then, as the universe cooled, dimensional bifurcation took place, with the 3+1-dimensional universe known to us today expanding differentially vis-à-vis the other dimensions, the latter being concealed from us. Corresponding to this dimensional divergence was a divergence of the forces of nature from their primordial condition of indistinguishability: the gravitational force "froze out" in the GUT era, the strong force in the electroweak era, and the weak force in our present era (where electromagnetism holds sway). To this extent then, mainstream and phenomenological views of cosmogony are largely in accord. We know, however, that there is a fundamental disparity between these approaches, one that we have framed via the issue of symmetry.

The phenomenological approach to cosmogony does not begin with the notion of "perfect symmetry" and assume that the universe evolved through a process of symmetry breaking. The starting point instead is an archaic epoch governed by the principle of "neither/nor": neither symmetry nor asymmetry. The "quantum soup" represented in stage 1 of Table 10.1 thus does not consist of objectively existing, already well-formed and self-symmetric particles in random relation to one another; rather, it entails a chaotic state of affairs in which particles — or spatio-sub-objective braneworlds — are lacking in symmetry. In stage 1, dimensional particles are indistinguishable not because of being symmetrically interchangeable but because of being largely undifferentiated from one another. Then, in subsequent stages, dimensional generation ensues in a process of Individuation wherein symmetry is first created.

So I am suggesting that, phenomenologically, cosmogonic symmetry transformation does not entail a lessening of symmetry, as in the conventional view, but the development of it. While this is true, the analysis provided in the previous chapter (Section 9.2) leads us to the conclusion that, in the two approaches, we are in fact dealing with transformations of *different kinds* of symmetry. Again, the Planck era symmetry of nature's forces alleged by convention alludes to a statistical interchangeability of individual particles, with the particles also presumed to be perfectly self-symmetrical at their cores. Symmetry is thus assumed with respect to both the external relationships among particles and to the core integrity of the particles: fundamental particles are taken to be indivisible from the first, completely self-contained individuals. For such idealized atomic entities, no internal individuation process is conceivable. On this account, only the external order of symmetry is purportedly transformed by being "broken."

From the contrasting phenomenological standpoint, the Planck era is characterized by a pervasive *absence* of symmetry. Here there can be no relation of external symmetry among particles since the particles are not symmetrical within themselves; being embryonic or fractional, they lack an integral core. They are therefore wide open to each other — not in the sense of the integration of internally well-bounded entities, but in the sense of their lack of differentiation. As noted in Chapter 9, the primordial fusion of force particles was in fact a *con*fusion, so that instead of having a symmetrical exchange of internally well-defined particles, there was a pre-(a)symmetrical flux in which particles had no definite internal values, in which there were no self-identical particles, no "ultimate atoms," no individuality. From this initially inchoate state of affairs, symmetry is generated in the process of Individuation whereby the dimensional particles become internally cohesive and well differentiated from each other. The process of symmetry making I am speaking of does not concern statistical generalities or abstract mathematical symmetry groups. It deals instead with the formation of the concrete individual, the actual particle event constituting a braneworld (lifeworld, dimensional organism). It is this process that is wholly obscured in mainstream cosmology's idealization of particle self-symmetry. Remember too that the Individuation process is essentially cyclogenic in nature: it entails the generation of spin, of topodimensional vortex waves. In sum, whereas external symmetry is allegedly transformed in the conventional approach by being "broken," in the topo-phenomenological approach symmetry transformation is understood as the generation of intrinsic topological spin taking place within the concrete individual event.

Now, it has indeed proven difficult for contemporary physicists to definitively specify the dynamics of the very early universe. Numerous speculative models of expansion and inflation exist, with many analysts departing from the standard view of a single inflationary era to propose several of these (e.g., Linde 2005; Jeannerot, Rocher, and Sakellariadou 2003). Inflation is related to force-field symmetry breaking. As astrophysicist John Gribbin (2000) puts it: "The key point is that the splitting apart of the components of the original grand unified force would have been associated with *scalar fields* that acted as a kind of antigravity, pouring energy (essentially, some of the original mass-energy) into the [inflationary] expansion of the Universe" (pp. 308–309). In Table 10.1 we envision three stages of inflationary expansion, each phase transition being associated with its own *internal* symmetry transformation event.

Bringing in the Kaluza-Klein aspect, inflation is also related to the dimensional bifurcation process wherein micro-scaled dimensions that were together in the Planck era subsequently split off from one another, with certain dimensions expanding relative to the others and the still compact dimensions becoming obscured when considered from the standpoint of the expanded dimensions. Because dimensional events are not just spatial abstractions in our spatio-sub-objective account, Kaluza-Klein theory is incorporated quite naturally: the dimensional transformations *are* the fundamental particle events, with each bosonic particle constituting a space-time braneworld. The three clockwise phase transitions of Table 10.1 that specify inflationary episodes are thus inherently dimensional. Here, dimensional bifurcation is not merely characterized as a "spontaneous" happening for which no real explanation is forthcoming; rather, it is understood as an integral feature of the process by which dimensional particles Individuate. Since we will explore this in detail in Sections 10.4 and 10.5, presently I will limit myself to offering an indication of the general pattern of Individuation.

In the opening stage, the three non-gravitational braneworlds or "branewaves" reside as embryos within the zero-dimensional matrix of the gravitational carrier wave (G). Advancing to stage 2, the higher dimensional waves expand, gaining their measures of cyclonic self-symmetry, of autonomy or Individuation. With this phase transition, the dimensional matrix is reduced as zero-dimensional gravitation decouples, "freezing out." Whereas gravitation presided as carrier wave in stage 1, when it is detached from the interrelational matrix in stage 2 it is partially eclipsed, persisting as but a penumbra. It is the one-dimensional strong force (g) that holds sway in stage 2. Having attained its (quasi-)Individuation, it now serves as carrier wave to the still embryonic weak and electromagnetic branewaves (g/(W, Z) and g/γ, respectively).

Two more inflationary bifurcations occur in the clockwise stages of cosmogony, two additional force particle decouplings and concomitant matrix reductions: in stage 3, the strong force freezes out and the weak force (W, Z) expands to autonomy, now serving as carrier for the still fractional electromagnetic force ((W, Z)/γ); in stage 4, the weak force freezes out and electromagnetism matures. The overall course of events attendant to dimensional expansion and divergence might be summarized, at least in part, by the observation that branewaves attain their symmetry and serve as carrier waves to still-developing, not-yet-symmetric branewaves. Carrier waves then freeze out and are eclipsed as higher dimensional branewaves realize their own internal symmetry or Individuation. However, the zero-dimensional gravitational vortex, being "selfless" (Chapter 7), has no symmetric locus of its own and thus serves purely as a carrier wave for higher dimensional waves. It seems we must also say that the three-dimensional electromagnetic wave, in attaining its symmetry, does not play the role of carrier since there are no still-developing vortices of higher dimension than it (at least not in the three-dimensional winding of the cosmodimensional spiral; see below).

For the stages of Individuation we have been speaking of, we only have *quasi*-Individuation. Our branewaves do not achieve genuine maturity before "reversing gears," switching from clockwise, projective action to counterclockwise, proprioceptive action; from dimensional divergence and expansion to a dimensional convergence and contraction in which earlier "movements" of the cosmic "symphony" are reprised (see Section 10.5). In this regard, while most cosmologists now appear to agree that the early universe was characterized by phases of expansion, there is less agreement on the fate of the universe. Will it continue to expand, or will it reverse itself and begin to contract? Paul Steinhardt and Neil Turok (2002) have addressed this question by offering a new rendition of the oscillating universe idea, one that has attracted much recent attention among astrophysicists. Taking their cue from string/M-theory, Steinhardt and Turok propose a cyclic model in which the universe undergoes countless rounds of expansion and contraction. What our topo-phenomenological version of string theory implies is a self-evolving cosmos that is not merely cyclical but *spiralic*. Agreeing with Steinhardt and Turok, I suggest that the cosmic expansion we are currently experiencing will be followed by a contraction, after which another period of expansion shall ensue. But the subsequent expansion will not just repeat the previous one. It will involve the opening up of a whole new dimension, including new forms of matter and a new force of nature beyond what are shown in Table 10.1 (it is in this novel winding of the dimensional spiral that the electromagnetic force will play the role of carrier wave). I shall have a little more to say about this prospect before I am finished. What I presently want to emphasize is that the overall picture of cosmic development I am suggesting is neither of a simply open universe whose given dimensions expand indefinitely, nor of a closed universe featuring endless cycles of expansion and contraction. We must imagine instead a self-evolving cosmos whose contractions are the "labor pains" that accompany the birthing of new dimensional organisms. Each cosmic organism expands toward Individuation in its turn, only to experience cosmic contractions that complete its Individuation in earnest and prepare it for the next round of creative cosmic growth.

Dare I say that the topo-phenomenological approach provides better grounding for cosmogony than does the consensus view? To be sure, conventional thinking about the evolution of the universe is rooted in centuries of sound mathematical theorizing and empirical research. But the up-shot of what we have seen in previous chapters is that the dominant paradigm of logico-empiricism is hard put to deal with the paradoxical facts of contemporary physics and cosmology. Therefore, in the absence of a coherent theoretical and philosophical framework, physicists operating within the orthodox paradigm have been obliged to resort to free-floating abstractions that defy direct observation, to self-serving ad hoc procedures and patchwork stratagems. As of this writing, physicists continue to cast about for the "right theory," when what is actually required is not just a new theory but a new *intuition*, one that surpasses the intuition of object-in-space-before-subject that has undergirded the scientific enterprise from its inception. The crux of what I am proposing is that *phenomenological* intuition can provide what is necessary.

If linking the evolution of the universe to transformations of consciousness seems odd, it is because we have been deeply conditioned to regard cosmological events in purely "objective" terms, as physical occurrences "out there in space," with subjective awareness playing no role. But phenomenological intuition guides us to a different sensibility. From this standpoint, cosmogony cannot properly be understood as a sequence of happenings devoid of inner life, of lived subjectivity. Instead we must regard it as a process of Individuation that involves the self-transformations of sentient dimensional organisms (lifeworlds, braneworlds). In the spatio-sub-objective context, we are able to appreciate that the objective expansion of the universe is at bottom inseparable from a subjective expansion or inflation in which awareness of origins is occluded. By the same token, we can grasp the idea that the "reversal of gears" does not just bring about a physical contraction of the cosmos but one that is *psycho*physical, one in which dimensional organisms are transformed by gaining cognizance of their cosmic beginnings.

10.3 The Role of the Fermions in Dimensional Generation

The gloss of cosmogony provided above must be followed up with a more detailed study that examines cosmic history one stage at a time. But let us first consider the role of a key player in this story of creation: the matter particle or *fermion*.

In Chapter 7, we found that dimensional generation proceeds in such a way that enantiomorphic fusion results in the distillation of dimensional bounding elements, these elements crystallizing from the introjection of lower-dimensional carrier waves (the elements are given within square brackets in Table 7.1). In this earlier chapter, I intimated the *dialectical* nature of depth-dimensional bounding elements. Whereas classical bounding elements are simply self-symmetric entities that are juxtaposed without overlapping (like separate beads on a string), dialectical bounding elements, being essentially relational, are distilled in oppositional pairs as asymmetric enantiomorphs. The intra-dimensional enantiomorphy of the bounding elements mirrors the enantiomorphic relationship between dimensional ratios prior to fusion (the enantiomorphic character of the bounding elements is not indicated in Table 7.1). Moreover, while the classical situation is one in which the bounding elements of space are strictly set off from space itself, the former being local and the latter global, in the depth-dimensional case we are dealing with "global localities" (Chapter 6), and this implies that synsymmetry is at play. In their local aspect, elements are related asymmetrically. But the local continually flows into the global uniting opposing elements to bring symmetry. Thus the "local" bounding elements of a dialectical space and the space itself are intimately related in a dynamic process whereby symmetry is made and broken in an ongoing way. I can think of no better visual model of this than Figure 9.3(b). Opposing perspectives of the cubic structure repeatedly merge to create continuity or symmetry, then decompose to bring discontinuity or asymmetry. And the two opposing actions do not merely occur in separate instants of time but paradoxically overlap and interpenetrate.

Now, we know that in the clockwise stages of dimensional generation, each phase transition brings a projection of object-in-space-before-subject. Whereas underlying dimensional action is concretely universal in nature, the objects that are projected appear as finite particulars, substantive things, material bodies and energies. The bodies are contained in space via the medium of space's bounding elements. For, to contain such an object, one must establish a boundary that sets off its interior region from what lies outside of it; whatever other properties it may possess, without an inside and outside there can be no object-in-space. We may say, then, that the bounding elements of dimensions are what set up the projected world of material objects, of matter-energy.

Again, lifeworld bounding elements are not just geometric abstractions. In Table 10.1, the dimensional bounding elements gain physical expression as the paired particles enclosed within square brackets. Whereas the unbracketed terms within the matrix cells denote gauge bosons, the bracketed particles correspond to the six pairs of fermions given in the standard model of particle physics. In the standard model, the fermions are arranged in three families that differ from one another only in terms of the masses of their members (Table 10.2). Particle families are also referred to as "generations," the generations with greater masses being associated with earlier stages of cosmogony. Consonant with the dialectically paired nature of the dimensional bounding elements in our depth-dimensional account, each family of the standard model contains two pairs of opposing particles. A lepton pair consists of an electron or more massive electron counterpart (muon or tau), and a corresponding neutrino. Quark pairs consist of opposing quarks: up-down, strange-charm, and bottom-top. Opposition between the members of fermion pairs is expressible via a technical property known as "isotopic spin." In terms of isospin, pair members are distinguished by the fact that one "spins up" (possessing a spin value of $+1/2$) while the other "spins down" (with a value of $-1/2$).[4]

	I	**II**	**III**
Lepton Pairs	electron (e) .00054	muon (μ) .11	tau (τ) 1.9
	electron neutrino (v_e) $<10^{-8}$	muon neutrino (v_μ) $<.0003$	tau neutrino (v_τ) $<.033$
Quark Pairs	up (u) .0047	strange (s) .16	bottom (b) 5.2
	down (d) .0074	charm (c) 1.6	top (t) 189

Table 10.2. The three families of matter in the standard model. The particles are shown with their associated masses, given in multiples of the proton mass.
(Adapted from Greene 1999, p. 9.)

The fermions are indeed fundamental particles. They account for all the matter-energy of which our universe is composed, while they themselves are not divisible into smaller units of particulate matter. In our fully realized Kaluza-Klein rendering of the standard model, we can go beyond simply noting the "atomic" or irreducible nature of the fermions. Here, where every particle has its dimensional role, the role played by the fermion is that of dimensional bounding element. The fermions lend substance to bosonic dimensionality. They ground the bosons, serving to mediate between these largely immaterial particles and the world of matter and energy. (While most gauge bosons are massless, the weak variety do appear to acquire mass in the course of cosmogony.)

The bosons themselves are of course also fundamental entities, *more* fundamental, in fact, than the fermions, since fermions derive from the fusions of enantiomorphically-related boson ratios. We may say that whereas fermions, as bounding elements, embody the local aspect of a given braneworld, bosons express the global aspect of the "global locality," i.e., the braneworld as a whole. The close dialectical relationship between bosons and fermions is not understood in conventional thinking about fundamental particles. In fact, relatively little is actually understood here about the underlying relationship between bosons and fermions. In an effort to unite force and matter particles in one overarching symmetry, the abstract notion of "supersymmetry" was put forward in the 1970s. This concept has been highly influential in contemporary physics but it is far from parsimonious, since the formal unity it confers requires that each of the known particles be accompanied by mysterious "superpartners" of opposing type (the partners of the known fermions are hypothetical bosons, and the converse). While the supersymmetry idea thus doubles the number of particles taken as fundamental to nature, there is as yet no evidence that would support the actual existence of the alleged superpartners.

In the contrasting topo-phenomenological approach to boson-fermion relatedness, no such proliferation of particles is necessary, for the known fermions are seen as directly evolving from the interplay of primordial forms of known bosons. The proposed primary status of the gauge bosons is generally consistent with Penrose's view that "massless particles and fields...be regarded as more fundamental than massive ones [like the fermions]," and "that massive particles and fields should probably be, in some sense, built up from massless ones," with the latter playing "some sort of primitive role" (1987). The basic idea is further articulated by Englert, Houart, and Taormina (2002). Their formal analysis leads them to the conclusion that space-time fermions are distilled from the dimensional "compactification" and "truncation" of primordial bosonic structures with nonorientable topology (nonorientable structures like the Moebius strip and Klein bottle are seen as playing significant roles in the process). The authors close their paper by noting that the "emergence of the space-time fermions from diagonal subgroups of direct products of space-time groups and internal groups is...suggestive of the generation of fermions through excitations from topological boson backgrounds" (p. 14).

Now, in Chapter 5, the several eigenvalues of the Table 5.2 spin matrix were associated with primary forms of movement, dimensional waveforms capable of carrying signals or information that, in turn, can influence the behavior of matter and energy (Bohm 1980, pp. 150–51). In quantum mechanics, the information affecting the wave function for a matter particle can be said to be imparted by a non-local wave field that Bohm referred to as a "quantum potential" (see Chapter 4). We can therefore say that a dimensional wave's informative effect upon matter is mediated by a quantum potential. And what I now propose is that quantum potentials governing the actions of particular fermions are distilled from the annihilative fusions of particular enantiomorphically related bosonic waves. Thus, each fermion pair that arises in the course of cosmogony is associated with a distinctive quantum potential that guides its behavior. So the emergence of fermions from the interplay of primordial bosons is linked to the generation of quantum potentials that influence the action of the fermions. All in all, fermions and bosons are related reciprocally: the bosons govern the behavior of the fermions by imparting information to them, and the fermions ground the bosons by connecting them with the material world.

Doubtless more could be done by way of making the phenomenological rendition of particle physics answerable to the specific findings of this field, and more *will* be done before I am finished. But I must acknowledge the general limitation of my account in this regard. The shortcoming is necessitated by the purpose and restricted scope of this book. Assuming that the challenges confronting present-day physics cannot be met while continuing to operate within the currently dominant paradigm, my chief aim has been to outline a qualitatively different approach, a phenomenological alternative that speaks to the root presuppositions of conventional thinking about matter and energy, space and time. My hope is that the ideas offered here will stimulate enough interest that others will be motivated to join me in future efforts to further attune phenomenological intuition with the facts of modern physics.

10.4 Projective Stages of Cosmogony: Dimensional Divergence

Having considered the stages of cosmogony in broad outline in Section 10.2, let us now go back over them for a closer look. We refocus our attention on Table 10.1 for a more detailed understanding of the evolution of nature's forces and matter.

10.4.1 Stage 1

In this initial stage of cosmic evolution, we find the zero-dimensional graviton on the main diagonal, accompanied by three enantiomorphically related pairs of boson ratios in off-diagonal cells. G is the "fundamental tone" and the boson ratio pairs are "overtone-undertone" couplings. The overtones of the gravitational carrier wave support or contain the undeveloped, undertone forms of the other topodimensional gauge bosons. Thus, the g/G gravitational overtone, which is associated with the sub-lemniscate, carries the G/g strong nuclear undertone, the latter constituting the fractional or embryonic form of the one-dimensional lemniscatory gluon (g). Similarly, the (W, Z)/G gravitational overtone carries the weak nuclear undertone, G/(W, Z), which is a fractional form of the two-dimensional Moebial boson. Finally, the γ/G gravitational overtone carries the G/γ electromagnetic undertone, the latter being an embryonic form of the three-dimensional Kleinian photon.

The synsymmetric vortical events of stage 1 are indeed no more than embryonic. At this juncture, the symmetric and asymmetric aspects of the topodimensional branewaves are not clearly fused but *con*fused. Here the space-time fields are neither closed nor open, neither bounded nor unbounded. This largely unoriented state of affairs is what underlies the turbulent "quantum soup" prevailing just prior to the Planck time. To repeat, the Planck era chaos is more than just statistical. It is not simply that the basic force particles or branewaves are internally well-differentiated entities that are statistically interchangeable; rather, they lack differentiation, so that chaos holds sway at their core — something that Table 10.1, as a linear form of representation, obviously fails to reflect. In the Table, each dimensional particle is set off from all others by being enclosed in its own well-bounded cell. How different from this is the actual situation that prevails in stage 1. Despite the impression of simple boundedness conveyed by the compartmentalized layout of the Table, topodimensional boundaries are in fact nascent, seed-like, unformed. And yet, a certain kind of primordial order can in fact be found.

It is true that no *actual* difference exists among the branes in the initial stage of cosmogony. We could not even speak of a difference in developmental *potential*, if that word were meant in the positive, Aristotelian sense of an already existing capacity just waiting to unfold. Yet I have made the claim in previous chapters that dimensional spinors differ in their developmental potential from the outset. It is time to examine this idea more closely.

I propose that, in the opening stage, topodimensional branes are *implied*. I use this term in the special sense brought out by philosopher Eugene Gendlin (1991). Gendlin offers the example of a poet searching for just the right words to effectively express the next line of a poem. The poet appeals to a place in her body that Gendlin calls a "blank" or a "slot," signified by "....." (p. 61). In the ....., the next line of the poem is *implied*: "This ..... demands and implies a new phrase that has not yet come" (p. 61). "Yes, the next line is *implied*," says Gendlin, "although it does not exist and never has" (p. 63). By way of explaining further, Gendlin distinguishes the conventional meaning of "implicit" from his own:

> What we have once thought explicitly can become implicit when we stop thinking about it. Much previous thought is also implicit; it has been built into our situations and our lives although we have never thought it explicitly ourselves. But in our instance (of a poet), what *was* implicit can be new to the world. Let it stand that something quite new can have been implicit. (1991, p. 81)

Gendlin's concept of *implying* entails an unconventional way of thinking about time.[5] For an event occurring at time T_2 truly to be new, yet also to have truly been implied at T_1, time must work *retroactively*:

> Implying can imply something new of which we *then* say that it *was* implied. This is not wrong; rather this temporal relation works backward into time....The ..... did not contain the line. Now, from the arrived line backwards, the blank *was* the implying of that line. But it is not an error....It is not a confusion of memory. (p. 81)

It seems Gendlin's subtle concept of "implying" is well suited to the primal negativity of the Planck era matrix. By stating that the eigenvortices were "implied" in this nascent stage (S1), we surely do not say that the primordial branes existed in any well-differentiated, positive sense: "The ..... did not contain the line." And yet it is clear that we cannot just state categorically that these branewaves did *not* exist in S1, since, from the retroactive vantage point of subsequent stages, the differentiated branes that emerge *were implied* in S1 ("from the arrived line backwards, the blank *was* the implying of that line").

At one point, Gendlin associates the notion of "implying" with being "pregnant" (p. 75). Such a term accords well with the present approach, given that, in this book, we are essentially dealing with topodimensional processes of organic growth. We may surely say then that primordial branes are embryonic. Understood by the logic of neither/nor appropriate to the incipient phase of dimensional generation (see Chapter 6), an "embryo" is clearly not an Aristotelian potentiality. It is not a pre-formed, miniature version of the mature being, one that is "in there" already and just waiting to come out. Nor is the embryo a *mere* negativity, a state of simple and total formlessness from which any arbitrary form could be created, the creation having to be *ex nihilo*. Rather, an embryo is an *implication*.

10.4.2 Stage 2

With the passage to stage 2, the divergence of branewaves commences. Through the process of enantiomorphic fusion, the three non-gravitational waves expand toward maturation, while the gravitational wave "freezes out" in such a way that it is partially eclipsed, appearing in this one-dimensional milieu as a penumbral attenuation of its original, zero-dimensional wave form (although G is eliminated from the matrix in passing to stage 2, if we read the stage matrices of Table 10.1 as *overlapping* one another as discussed in Section 10.2, we may indeed infer that G persists as a penumbra). In completing the merger of enantiomorphs, stage-1 boson ratios are annihilated as free-standing entities. The gravitational overtone (g/G), having fulfilled its stage-1 role as carrier wave for the strong force undertone (G/g), is absorbed into the one-dimensional vortex, which has now surpassed its embryonic status and been brought to term as the quasi-mature gluon, g. In like manner, the weak and electromagnetic gauge bosons advance toward maturity via annihilative fusions of their corresponding enantiomorphic couplings ((W, Z)/G fuses with G/(W, Z), and γ/G with G/γ, respectively). It is from the introjection of carrier waves occurring in this process that the fermion pairs are distilled. The introjection of the g/G overtone produces the top and bottom quarks; the absorption of the (W, Z)/G overtone produces the tau and tau neutrino (τ and ν_τ); and the introjection of the γ/G overtone yields the charm and strange quarks. In this way, the support for the higher-dimensional bosonic waves provided by the zero-dimensional carrier wave in stage 1 is now internalized to the higher-dimensional waves, taking the form of fermionic bounding elements that serve to ground the bosons, to put them in relation to the material world.

On what basis do we associate particular fermion pairs with particular dimensional overtones? In the left-hand column of Table 10.3, overtones are ranked in ascending order of their dimension numbers, using the dimensional notation employed in Table 7.1 (whose values are the dimensional counterparts of the particle values given in Table 10.1). Here lower-dimensional values express more primordial spin states, as the stagewise layout of Table 7.1 suggests (remember that the native dimensionality of any dimensional ratio is determined by the value of the denominator). In the case of the fermions, we know from the findings of physics that primordiality is correlated with mass: the more massive the particle, the earlier is the period of cosmogony to which it is related. In the right-hand column of Table 10.3, fermion pairs are rank-ordered in terms of their masses (averaged over pair members). Fermion pairs and dimensional overtones are thus matched on the basis of their respective relative primordiality values.

Overtones, in ascending order of dimension number	Fermion pairs, rank-ordered by average mass (in multiples of the proton mass)	
	Fermion pair	Average mass
$\varepsilon_{D1}/\varepsilon_{D0}$	t, b	97.1
$\varepsilon_{D2}/\varepsilon_{D0}$	τ, ν_τ	.967
$\varepsilon_{D3}/\varepsilon_{D0}$	c, s	.88
$\varepsilon_{D2}/\varepsilon_{D1}$	μ, ν_μ	.056
$\varepsilon_{D3}/\varepsilon_{D1}$	d, u	.012
$\varepsilon_{D3}/\varepsilon_{D2}$	e, ν_e	.003

Table 10.3. Primordiality rankings of dimensional overtones and corresponding fermion pairs

Note that the quantities shown in Table 10.3 do not reflect the actual masses of the lower-dimensional GUT- and electroweak-era fermions appearing in Table 10.1; they only reflect the masses these fermions assume when they are detected and measured in the relatively low-energy regime of the three-dimensional laboratory. My suggestion is that we regard the fermions of the lower-dimensional stages of cosmogony as "hotter," more massive, primordial precursors of the counterparts observed in the laboratory. Again, when it comes to the conditions prevailing in the universe within 10^{-12} seconds of its origin, cosmologists have only been able to speculate. But it is widely believed that primordial forms of matter did exist that served as antecedents of the matter that we can observe. I propose that the GUT- and electroweak-era fermions given in Table 10.1 played a precursory role of this kind.

Let us now focus our attention on the stage-2 (S2) relationship among the gluon, weak gauge boson, and photon. In this stage, the first has reached quasi-maturity whereas the latter two continue to function in a fractional, embryonic manner. To be sure, the weak and electromagnetic braneworlds have gained a measure of Individuation insofar as they have disentangled themselves from the gravitational matrix. Yet they are now embedded in the matrix of the strong force. Unlike the utterly "selfless" graviton, the gluon is involved in its own process of self-development. Its relation to the weak boson and photon is therefore not purely supportive. Functioning as carrier wave, the S2 strong force does selflessly contain weak and electromagnetic self-containment. However, in the self-generative aspect of its functioning, the gluon also *threatens* that higher-dimensional self-containment. Soon we will see what this means, but let us first consider more closely the gluon's role as carrier wave.

The impression is conveyed from the layout of Table 7.1(b) that before the lemniscatory dimensional wave can serve as carrier to the Moebial and Kleinian waves, it must complete its own development. Only then does the one-dimensional wave feature the overtones ($\varepsilon_{D2}/\varepsilon_{D1}$ and $\varepsilon_{D3}/\varepsilon_{D1}$) that are necessary for the negative (selfless) containment of the two- and three-dimensional undertones ($\varepsilon_{D1}/\varepsilon_{D2}$ and $\varepsilon_{D1}/\varepsilon_{D3}$, respectively). Stated in physical terms, it seems the one-dimensional gluon must itself be fully Individuated before it can support the Individuation of the two-dimensional weak boson and three-dimensional photon. But I have acknowledged that Table 7.1(b) separates dimensional windings in an artificial fashion that neglects their temporal overlap. Tables 7.1(a) and 10.1 address this deficiency by showing cosmodimensional stages as interwoven. Still, while there is indeed temporal overlap among the several dimensional windings, we know that they are not simply contemporaneous, since each gyre possesses its own "time scale" (Chapter 7), its own order of temporality. How then are we to read the timing of the S2 gluon's selfless facilitation of weak boson and photon self-development? If this support does not simply come *after* the gluon's own cosmogony, or simply simultaneously with it (as Table 10.1 appears to suggest), when does it come?

The S2 functioning of the gluon as a carrier wave may be elucidated by turning to the theory of electromagnetism. Here we have the possibility of waves or energy potentials that are *advanced*. An "advanced potential" is defined as "any electromagnetic potential arising as a solution of the classical Maxwell field equations, analogous to a retarded potential solution, but lying on the future light cone of space-time; the potential appears, at present, to have no physical interpretation" (Lapedes 1978). According to philosopher of science Mary Hesse (1965), the application of advanced solutions of Maxwell's equations "leads to the apparently paradoxical result that one event affects another ... at a time ... *before* it occurs" (pp. 279–80). Applying this notion to the gluonic branewave's supportive containment of weak and electromagnetic waves, I propose that the gluon's influences are exerted as "advanced waves." What I am suggesting is that the one-dimensional gluonic vortex carries the higher-dimensional vortices, provides them with selfless support, at a time that is "before" the gluon has matured to the point of being equipped with overtones to do this! The overtone carrier waves by which the gluon assists the weak boson and photon in stage 2 are thus advanced waves.

However, is the notion of the "advanced wave" not an outright contradiction? How would it be possible for the gluon's supportive influence to be felt at a time that predates its attainment of the capacity to exert such an influence? It would *not* be possible if a single order of temporality prevailed in relations among dimensional waves. An effect cannot precede its cause on the same scale of time. The ability of the gluon to function "in advance" as a carrier of the weak boson and photon derives precisely from the fact that these dimensional waves constitute *different* orders of temporality. It is because the gluonic temporal gyre is more "tightly coiled" than its higher-dimensional counterparts that it can appear to be "ahead of itself" in relating to them.

Consider again the analogy of the clock introduced in Chapter 7. The relationship between the second hand and the minute hand provides a simple model of different time scales. All sixty ticks of the second hand occur within the same tick of the minute hand. Might we not say, then, that the faster, more compressed cycle is temporally undifferentiated relative to the slower time scale? Are not the distinct ticks of the second hand all the same tick to the minute hand, which cannot make such fine distinctions? In classical physics, of course, this relativistic effect is obviated by the fact that the two time scales are entirely commensurable; they are incorporated as inter-convertible subdivisions of a single metric. But what *we* are dealing with are not quantitative temporal differences but ontologically different orders of temporality that, by definition, are irreducible to a unitary time metric. The consequence is that relativistic effects cannot be eliminated. Since we are unable to reduce the temporalities of the strong, weak, and electromagnetic braneworlds to a common scale of measure, we must entertain the notion that, while different stages of Individuation are surely distinct within the relatively tight temporal coiling of the gluonic vortex, in relation to the more "loosely wound" circulations of the weak and electromagnetic vortices, the moments of gluonic cosmogony do lose their distinctiveness: all occur at the "same time." It is this "temporal compression" of the one-dimensional winding relative to higher-dimensional windings that allows the gluon to function as carrier wave in "advance," to negatively contain the Individuation of the weak boson and photon at the "same time" that it has not yet completed its own Individuation. In its latter "moment" of unfulfilled development, the gluon affects the higher-dimensional bosons in a manner that is contrary to the mature gluon's supportive influence. Instead of containing weak and electromagnetic self-containment, the gluon menaces it. How so?

To contain is to enclose within boundaries. The self-containment of weak and electromagnetic dimensions receives nothing but support (negative containment) within the matrix of the utterly boundless graviton (we may associate primal gravitational action with the vortical action of the *apeiron*, the "boundless giver of boundaries"; see Chapter 4). In the matrix of the gluon, however, the blessings are mixed. Insofar as the S2 gluon is not only operating "in advance" as a boundless carrier wave but also as a branewave still involved in its own self-binding, its eigenfunctioning poses a danger to the functioning of the higher-dimensional branewaves. Why is that so? In stage 2, the self-containment of the gluon has gone further than that of the weak boson and photon. The latter two are but weakly self-contained vis-à-vis the former; being still largely undifferentiated from the gluon, they are wide open to its influences. Those influences are "positive" in the quasi-mature moment of the gluon. Rather than deriving from the selflessly supportive aspect of gluonic functioning, they stem from the gluon's continuing efforts to contain itself. Therefore they threaten to flood the relatively fragile self-containment capacities of the embryonic weak boson and photon. In sum, the S2 gluon affects the weak boson and photon in both a "positive" and "negative" way: in its moment of quasi-maturity, the gluon's act of establishing its own identity, securing its own boundaries, can overpower the permeable boundaries of the less developed, higher-dimensional branewaves; in its moment of full maturity, when it can function as a carrier wave, the gluon selflessly supports the higher-dimensional waves. To complete the account of S2 gluonic functioning, let us spell out the gluon's relation to the graviton: with the passage from S1 and concomitant fusion of g/G and G/g enantiomorphs, the graviton freezes out in such a way that its g/G overtone is introjected by the developing gluon to yield top- and bottom-quark bounding elements for gluonic self-containment. In the frozen-out form which it takes in the one-dimensional milieu of the gluon, the graviton is an attenuated version of the original zero-dimensional graviton. The latter is now overshadowed, "positively contained" in the one-dimensional environment in the sense of being repressively "bottled up" there.

10.4.3 Stages 3 and 4

Developing braneworlds continue to diverge in the next two phase transitions. The passage to the third stage of projection is marked by enantiomorphic fusions through which weak and electromagnetic vortices achieve further maturation, while the strong-force vortex freezes out in a manner wherein its overtones are introjected and it is partially eclipsed, appearing in this two-dimensional environment as a shadow of its original, one-dimensional wave form. In completing the merger of enantiomorphs in this stage, the boson ratios of stage 2 are annihilated as free-standing entities. The gluonic overtone, $(W, Z)/g$, having discharged its stage-2 role as carrier wave for the weak force undertone, $g/(W, Z)$, is absorbed into the two-dimensional vortex, which has now transcended its embryonic status and been brought to term as the quasi-mature weak boson duo, W and Z. From the S3 introjection of the $(W, Z)/g$ carrier wave, the μ and ν_μ leptons are distilled as bounding elements for the weak boson. Similarly, the photon moves toward maturity via the annihilative fusion of γ/g and g/γ enantiomorphs. The S3 introjection of the γ/g carrier yields the up and down quarks for the still-developing fractional photon, $(W, Z)/\gamma$.

The stratification of functioning attendant to the third stage of cosmodimensional projection is such that the eclipse of the S2 gluonic matrix is accompanied by an even deeper overshadowing of the S1 gravitational matrix. Viewed in terms of the containment metaphor, we may say that occluded matrices are repressively contained or "bottled up" in such a way that "bottles within bottles" are created: the S2 matrix is "positively" contained within the S3 matrix, and the S1 matrix within the S2 matrix. It can also be said that, in S3, the photon is contained by the weak boson pair. In this case, however, the containment is *negative*. That is to say, the still immature (W, Z)/γ electromagnetic wave is not eclipsed by the weak-force wave but carried by it via its γ/(W, Z) overtone, which operates as an *advanced* wave. Yet it is also true that the S3 weak boson duo, in the quasi-mature moment of its functioning, *endangers* the self-containment of the photon. Because the photon is still embryonic, it is susceptible to being flooded by the potencies of the weak boson.

The divergence of topodimensional gyres reaches its apex in the last projective stage of cosmogony. Passage to stage 4 brings a concluding enantiomorphic fusion through which the quasi-maturation of the electromagnetic vortex is completed, with the weak force freezing out, its native two-dimensionality being overshadowed in the three-dimensional milieu. The weak-force overtone, γ/(W, Z), having fulfilled its stage-3 function as carrier wave for the electromagnetic undertone, (W, Z)/γ, is absorbed into the three-dimensional vortex. From this introjection of the carrier wave, the e and v_e leptons are distilled as bounding elements of the photon. Functioning has become further stratified in this stage. With the two-dimensional weak force now being "positively contained" within the three-dimensional environment, four nested matrices presently exist, the penumbra shading back from stage 4 to the dark recesses of stage 1.

10.5 Proprioceptive Stages of Cosmogony: Dimensional Convergence

The stratification of functioning in the projective stages of cosmogony attends the divergence of the topodimensional gyres. Dimensional organisms give birth to themselves, bring about their own Individuation, by extricating themselves from the lower-dimensional wombs. It is in this process that lower-dimensional matrices become backgrounded or "bottled up." In the layering of matrices that this involves, dimensional windings are partitioned from one another. Of course, these cosmogonic "nativities" only bring *quasi*-Individuation. To give birth to itself in earnest, each dimensional organism must undergo a second nativity, one requiring that it "shift gears," switch its orientation from forward to backward, clockwise to counterclockwise. Once the transition is made from the posture of projection to that of proprioception, the dimensional organism can proceed to move backward through its "birth canal." As it retraces its steps to its point of origin, the lower-dimensional matrices from which it arises are no longer repressively occluded but consciously recognized. In this way, the lower dimensions emerge from "positive containment" and the several dimensional spheres harmonically converge. A mode of "integral awareness" (Gebser 1985) is realized in which dimensional spheres that had become more or less opaque during stages of dimensional divergence now become diaphanous.[6] (The shift from stratified consciousness to integral consciousness can be grasped with topo-experiential precision via the perspectival integration of the Necker cube discussed above.)

Advancing through the stages of proprioception, the matrix reductions enacted in the process of projection are reversed. With matrices now expanding, the containment relations between lower- and higher-dimensional waves are transformed. The now-Individuated higher dimensionalities are surely not just contained by the lower dimensions as in early stages of higher-dimensional development, nor do the higher dimensions contain the lower dimensions in the repressive fashion operative during quasi-Individuation. Being genuinely Individuated, the higher dimensions presently contain the lower *dialectically*, by acknowledging lower-dimensional containment of *them*. The mutual containment of branewave dimensions is thus realized and they enter into harmony. However, the *asymmetry* that is involved here must be kept in mind. In any given inter-dimensional relation, the higher dimensional member contains the lower largely in the interest of higher-dimensional self-development, whereas the lower contains the higher in an essentially selfless way. Not that we can treat the respective containment contributions of synchronized dimensions as if they could simply be teased apart. The harmony between these dimensions is so close that the containment of one by the other flows unbrokenly into its converse.

We have now reached the point where a more comprehensive account of topodimensional convergence is required, one that goes further than the partial indications previously given. Let us rehearse the stages of counterclockwise co-evolution in greater detail.

We know that, in the Kleinian winding of the dimensional spiral, the initial shifting of gears to the backward orientation occurs in stage 5. This involves the proprioception of electromagnetism, and of the lepton pair that accompanies the photon, viz. the electron and electron neutrino. From the projective standpoint of stage 4, electromagnetism is approached like any other phenomenon: the attempt is made to objectify it. But we saw in Chapter 8 how this effort has been frustrated. Beginning with the research of Michelson and Morley, the phenomenon of light has proven stubbornly resistant to objectification. To reiterate the conclusion of Young (1976), while the conventional "scientist...likes to think of [a photon of] light as 'just another kind of particle,'...light is not an objective thing that can be investigated as can an ordinary object....Light is not seen; it is [the] seeing" (p. 11). This comment is consonant with Heidegger's phenomenological insight into light, his association of it with "presencing" or "presence as such," rather than just with "what is present" (1964/1977, p. 390). For Heidegger, "presence as such" bespeaks *Being*. Thus, when light or electromagnetism is understood phenomenologically, it is grasped as a whole dimension of dynamic life process, as an organic lifeworld unto itself. On this reckoning, the phenomenon of light is no mere object appearing before a detached subject but is an *inseparable blending of subject and object*. The dialectical apprehension of light intimated in Chapter 8 is the *proprioception* of light, an action that becomes possible when the transition is made from stage 4 to 5 in the Kleinian winding. In gaining a sense of the blending of subject and object, access is gained to the prereflective source of their reflective separation.

At the same time, the source apprehended proprioceptively in stage 5 is pre-*spatiotemporal*. In Chapter 3 we found that the location of the subject in Einstein-Minkowski space-time is the here-now origin of the light cones. It is from the subject's here-now that elsewhere and otherwhen stretch out to contain the objects projected before said subject. The proprioception of light is tantamount to retracting the light cones of Einsteinian space-time, reeling them back into the here-now. In this backward movement of light to its source, the internal circulation of subject and object is at once a circulation of here with elsewhere, of now with otherwhen, and of space and time themselves (see Chapter 3).

The "black hole" uncovered at the core of the here-now via proprioception can also be said to constitute the core of quantum mechanical action ($\hbar$). With the proprioception enacted in stage 5, no longer is pre-spatiotemporal Kleinian spin ($\varepsilon\hbar/2$) kept under wraps so as to maintain in abstraction the old formula of object-in-space-and-time-before-subject (see Chapter 4). The proprioceptive reversal of gears lifts the quarantine on quantized dimensional action, bringing full awareness of the vortically spinning electromagnetic braneworld that underlies our ordinary experience of space and time.

The stage-5 proprioception of the photon includes a proprioception of its fermionic concomitants: the electron and electron neutrino. In stage 4, these leptons are regarded as but free-standing sub-atomic particles having the ontological status of objects in space. What comes to light with proprioception is that the lepton pair forms the pre-spatiotemporal *bounding elements* of the photon, the elements critically involved in relating electromagnetic action to the world of matter and energy. It is now understood that the γ–e–v_e triad constitutes a "global locality," with the photon's prereflective global spin being closely accompanied by the local spinning of the electron–electron neutrino pair that grounds the photon in material reality.

In passing to the second stage of counterclockwise action, dimensional convergence begins. The matrix expansion attendant to this phase transition signifies that the electromagnetic organism — moving backward through its "birth canal" in a second self-nativity — is proprioceiving its stage-3 nascency, the stage wherein the then immature photon, (W, Z)/γ, had been flooded by the potencies of the quasi-mature weak boson pair, W and Z, yet had received negative containment from that lower-dimensional pair as well via the γ/(W, Z) advanced potential. The stage-6 reprise of this earlier situation is clearly no mere regression to it. For, with the new proprioception, the stage-5 realization of electromagnetic Individuation is not just left behind, nor does it merely persist as a background penumbra. The Individuation of the previous stage is wholly transparent in stage 6 so that the now mature photon is able to *contain* the weak boson's potencies rather than be overwhelmed by them. Included in the retrograde electromagnetic action of stage 6 is the proprioception of up and down quarks as bounding elements of pre-spatiotemporal electromagnetic spin.

For its part, the weak force has also undergone development. By its fourth stage of Individuation, it has advanced beyond quasi-maturity and now enacts its own proprioception, one entailing the self-apprehension of its two-dimensional pre-spatiotemporal spin, including its μ and ν_μ bounding elements. When completion of weak-force Individuation is taken in relation to the now-Individuated electromagnetic organism, dimensional organisms presently contain each other in the asymmetric fashion noted above, since the lower-dimensional organism completes its maturation by realizing selflessness vis-à-vis the higher-dimensional self. The Moebial and Kleinian vortices now gyrate in synchrony.

The harmony of dimensional spheres is surely not complete with the notes that have thus far been struck. Though the electromagnetic and weak forces have become Individuated with respect to each other, neither has yet become fully attuned to the strong and gravitational forces. The next group of counterclockwise circulations brings a further convergence of topodimensional spheres. With the matrix expanding once more, the electromagnetic organism presently proprioceives the even more primal, stage-2 situation in which — as the g/γ embryo, it had been overpowered by the potencies of the quasi-mature gluon, g, yet had also received negative containment from that one-dimensional boson via the γ/g advanced potential. Of course, the stage-7 photon is mature now and therefore can wholly *contain* the gluon's potency rather than be fractured by it. Included in the electromagnetic action of stage 7 is the proprioception of charm and strange quarks as bounding elements of prereflective electromagnetic spin. The gluon, for its part, has undergone its own Individuation. By its third stage of development, it has gone beyond the projective stage of quasi-maturity and currently enacts a proprioception of its one-dimensional spin, which includes its top and bottom quark bounding elements. Thus attaining Individuation in synchrony, electromagnetic and strong-force braneworlds become attuned to each other, containing each other in the asymmetric way. Kleinian and lemniscatory vortices spin together.

In the group of counterclockwise actions we are examining, it is not only electromagnetism that harmonically converges with the strong force, but the weak force as well. The latter dimensional organism, moving backward through its own "birth canal" in its fifth stage of Individuation, proprioceives the stage-2 state of affairs in which — as the $g/(W, Z)$ embryo, it had been inundated by the potencies of the gluon, yet had also received negative containment from it via the (W, Z)/g advanced potential (the S5 proprioception of weak bosonic spin includes the τ and v_τ bounding elements). The now-Individuated weak boson is mature and can fully contain the gluon, which reciprocates this containment in a selfless way, bringing asymmetric mutual containment of lemniscatory and Moebial braneworlds. Still, the harmony of worlds has not yet been carried to completion. Although electromagnetism and the strong and weak forces all move in synchrony with respect to each other, none are wholly attuned to gravitation.

The force of gravity enters the picture in the final group of counterclockwise circulations. With the full-blown expansion of the cosmogonic matrix, the wholly Individuated stage-8 photon proprioceives the negative containment of its stage-1 embryonic spin (G/γ) within the "black hole" of the primal gravitational field (γ/G). This happens in such a way that electromagnetic and gravitational braneworlds contain one another asymmetrically. Similarly, the fully mature, stage-6 weak boson proprioceives the gravitational containment ((W, Z)/G) of its embryonic spin (G/(W, Z)) in a manner that brings about the asymmetric mutual containment of weak and gravitational braneworlds. And lastly, the Individuated stage-4 gluon proprioceives the gravitational containment (g/G) of its nascent dimensional action (G/g) so as to achieve asymmetric mutual containment of strong and gravitational braneworlds. All topodimensional vortices — Kleinian, Moebial, lemniscatory, and sub-lemniscatory — are now turning in asymmetric synchrony.

Since branes are generalizations of *vibrating strings*, we may close this analysis on a musical note. Our primary bosons are "fundamental tones" in a dimensional symphony that reaches its crescendo with the achievement of their full-fledged harmony. This is the way physics' long-sought goal is realized. After a century of dissonance, the forces of nature are brought into tuneful accord.

10.6 Conclusion: Wider Horizons of Cosmic Evolution

I have suggested that the shape of the self-evolving cosmos is essentially that of a spiral. The process begins with the expansion of the universe through stages of cosmodimensional projection. This is accompanied by the occlusion of lower dimensions (as indicated by the reduction of the matrices; when the universe expands matrices contract). The "cosmic gears" then shift and the universe undergoes contraction, as proposed by Steinhardt and Turok in their cyclic model of cosmogony (see above, Section 10.2). Of course, from the phenomenological standpoint, we are not dealing here with merely objective physical events. The self-evolving cosmos is a *psycho*physical affair whose stages of contraction bring proprioceptive awareness of the previously eclipsed dimensionalities (the contraction of the universe is thus marked by the opening up of the matrices to disclose lower-dimensional braneworlds). And, because cosmogony is not just cyclical but spiralic, the new round of expansion that follows the contraction does not simply repeat what went before.

Therefore: when the "birth contractions" of the dimensional organisms have been completed in the present round of cosmic evolution; when the four forces of nature have come into full synchrony; when the four branewaves are oscillating in close harmony — the dimensional spiral will then expand in logarithmic fashion, with the dialectical matrices now growing from 4^2 to 5^2. In this wider turning of the spiral, a new and more dialectically intricate, *four*-dimensional braneworld will come into play beyond the Kleinian world, a "meta-Kleinian" topological action pattern laid out in 5 × 5 matrices. In this novel world, new forms of matter and a new force of nature will emerge. The challenge of Individuation will be extended once more in the epoch to come, and the music of the dimensional spheres will resound.

Notes

1. Whitehead's "actual occasion" is a fruitful concept and the physicist Henry Stapp (1979) attempted to adapt it for use in quantum mechanics. As far as I can tell, however, Whitehead made no sustained attempt to explore the internal topology of the actual occasion. By default, this left the two basic modes of the occasion — the continuous or objective mode and the "atomic" or subjective mode — symmetric within themselves and externally related to one another. In the topo-dialectical alternative that I am offering, the internal structure of the actual event is not symmetric but dynamically synsymmetric. See also my related comments on Whitehead in Rosen 1994.

2. The term "archetype" is understood here not as a static Platonic form but a first *action*, from the Greek *arche*, the beginning, the first, and *typos*, "a blow, the mark of a

blow...from *typtein*, to beat, strike" (*Webster's New Collegiate Dictionary*, 1975 ed., s.v. "archetype").

3. The rendition of cosmogony provided by Table 10.1 is not entirely comprehensive. It does not take into account the question of *antimatter*. Without going into this issue in depth, we may assume for simplicity's sake that each fermion shown in the table has its antimatter counterpart. Cosmogonically speaking, we may surmise that particles were initially undifferentiated from their antiparticles. Then, as the universe cooled, differentiation occurred to the limited extent that particles and antiparticles appeared fleetingly in unstable pairs. Finally, with the further cooling of the universe, the anti-matter counterparts of matter were overshadowed, leaving the appearance of matter particles alone. Beyond this conjecture, I will say no more on the question of antimatter, lest the scope of this book become unwieldy.

4. It is worth emphasizing that while fermions are indeed grouped in pairs in the standard model, pair members are not internally related to each other in the deeply dialectical manner of the depth dimension. Rather than being understood as enantiomorphs that are synsymmetrically bound together, each pair member is taken as simply symmetric within itself, thus essentially divided from its counterpart. It is the underlying presupposition of symmetry that has made it so difficult for physicists to account for the chirality (mirror asymmetry) of the universe suggested by the phenomenon of parity violation in weak interactions. While I believe this important finding could readily be accommodated in the topo-phenomenological alternative to the standard model, I will refrain from further elaboration here (for a topodimensional discussion of parity violation, see Rosen 1994).

5. See Gendlin and Lemke's (1983) critique of the conventional notion of linear time as it operates in contemporary physics.

6. The themes of dimensional transformation and integral consciousness were explored extensively in a different context by the cultural philosopher Jean Gebser (1985). For more on Gebser, see next chapter.

Chapter 11
The Psychophysics of Cosmogony

Our psyche is set up in accord with the structure of the universe, and what happens in the macrocosm likewise happens in the infinitesimal and most subjective reaches of the psyche.

C. G. Jung, *Memories, Dreams, Reflections*

11.1 Psychical Aspects of the Fundamental Particles

Let me say again that the fields and particles of contemporary physics are not *merely* physical entities, but psychophysical in nature; at bottom they are dimensional actions implicating different orders of subjectivity. Until now, however, I have done little to specify the subjective aspects of the particles. The time has arrived for spelling this out.

In *Topologies of the Flesh* (2006), I applied my topodimensional analysis to psychological functioning. That work was carried out independently of any considerations arising in physics and cosmogony, yet its structures and patterns of development closely mirror what is given here. A total of sixteen basic psychological functions were identified, with four "eigenfunctions" (i.e., self-functions serving as primary elements[1]) and twelve inter-dimensional linking functions. The functions were seen to evolve in accordance with the same spiral logarithmic pattern of dimensional matrix transformation as found in the present work. From the dualistic viewpoint of classical thinking, it might seem astonishing that a detailed dimensional analysis of the psyche would correspond so closely to an analysis of particle physics. The discovery is less surprising from the phenomenological standpoint. In this context, the finding serves as confirmation that dimensional action is indeed *psychophysical*, that the fundamental physical structures constitutive of the self-evolving cosmos possess psychical aspects.

The psycho-functional analysis performed in the earlier volume is too intricate and extensive to repeat in its entirety. For our present purposes, I will limit myself to an abbreviated, summary account. Readers who wish to study in detail the basis of my conclusions about the psychological functions and their dimensionality can consult Chapters 6 and 7 of *Topologies of the Flesh*.

In *Psychological Types* (1971), the psychologist C. G. Jung maintained that the functioning of the human psyche entails four basic forms of activity. These functions cannot be reduced to each other and are invariant relative to the specific content of experience, which changes from moment to moment. The four functions of which Jung spoke are *thinking, feeling, sensing*, and *intuition*. Jung viewed thinking and feeling as diametrically opposed to each other and as "rational." That is, both are *representational* functions; they involve a reflection on, and an (e)valuation of, externally experienced reality from the inner perspective of the subject. In the case of thinking, the subjective base is abstract and disembodied (e.g., discursive reasoning, logical deduction, mathematical calculation), while, in the case of feeling, it is more concrete and embodied (I value what lies outside me in terms of the likes and dislikes of my body).

Beyond these "rational" operations lie the "irrational" functions of sensing and intuition, also seen to make up a pair of opposites. Irrational activity can be characterized as more "presentational" than representational; that is, it entails a more immediate (less reflective) reaching out to, grasping, and taking in of that which is other in the field of experience. In sensory perception, we experience the world discretely, dividing the objects we encounter into units that are bounded from each other. On the other hand, intuiting the world means apprehending it as an undivided whole, as in cases of "hunches" or "visions," where we are seized by a nebulous impression about the general course of events but are unable to account for the source of the presentiment. Note that, while the irrational functions are attenuated in adulthood, being constricted by the ascendant influence of rational thinking, according to Jung these functions originate in an "infantile and primitive psychology" that serves as the "matrix out of which thinking and feeling develop as rational functions" (Jung 1971, p. 454). Evidently, intuition would be the *most* primal form of experience, for while sensation involves "perception via conscious sensory functions" (p. 538), intuition entails "perception by way of the unconscious" (p. 518), through "dreams and fantasies" (p. 539), through a mode of operating in which one surrenders oneself "wholly to the lure of possibilities" (p. 519).

What I demonstrated in *Topologies of the Flesh* is that Jung's four basic functions correlate with the four orders of the lifeworld that emerge in the topodimensional analysis (Merleau-Ponty alternatively spoke of orders of the "flesh," hence the title of that book). Thinking is associated with the three-dimensional Kleinian realm that, in the present volume, we have related to electromagnetism or light. In fact, though *Topologies* does not deal with the force fields of physics, it does touch on the intimate connection between thinking, light, and three-dimensional Being. Heidegger is paraphrased in this regard: "to think Being is to think light" (Rosen 2006, p. 261).

The association in *Topologies* of the more concrete feeling function with two-dimensional Moebial action allows us presently to suggest that feeling or emotion constitutes the subjective aspect of the weak nuclear force. On first consideration, such a conclusion may seem baseless and odd. The persistence of conventional thinking might lead us to dismiss any possible linkage between a subjective quality like emotion and an objectively measurable physical entity like the weak force. But we must keep in mind that, from the standpoint of topodimensional phenomenology, the weak force is at bottom no mere physical object in three-dimensional space but a psychophysical lifeworld unto itself, one consisting of two dimensions. In our familiar three-dimensional world, thinking is the pre-eminent psychological function, with emotion tending to be overshadowed and repressed. The psycho-dimensional analysis provided in *Topologies* makes it clear that, in the two-dimensional lifeworld, it is feeling that holds sway. Then feeling must constitute the psychic quality of the two-dimensional weak-force braneworld. On a similar basis, the more primal psychological function of sensation is seen to be the subjective or psychic aspect of the one-dimensional lemniscatory braneworld where the strong force presides. Finally, intuition, the most primordial of the psychological functions, is the subjective aspect of the zero-dimensional, sub-lemniscatory gravitational world.

The analysis provided in *Topologies* is not limited to four psychological functions, since each function is seen to come in four forms. The resulting 4 × 4 matrix is given in Table 11.1, which — along with the functions, displays corresponding dimensional values and particle counterparts.

Thinking	**Feeling**	**Sensing**	**Intuition**
Rational ε_{D3} γ	Mental Emotion $\varepsilon_{D3}/\varepsilon_{D2} \rightarrow \varepsilon_{DB2}$ $\gamma/(W, Z) \rightarrow e, \nu_e$	Vision $\varepsilon_{D3}/\varepsilon_{D1} \rightarrow \varepsilon_{DB1}$ $\gamma/g \rightarrow u, d$	Mental Intuition $\varepsilon_{D3}/\varepsilon_{D0} \rightarrow \varepsilon_{DB0}$ $\gamma/G \rightarrow c, s$
Mythic $\varepsilon_{D2}/\varepsilon_{D3}$ $(W, Z)/\gamma$	Love ε_{D2} W, Z	Hearing $\varepsilon_{D2}/\varepsilon_{D1} \rightarrow \varepsilon_{DB1}$ $(W, Z)/g \rightarrow \mu, \nu_\mu$	Emotional Intuition $\varepsilon_{D2}/\varepsilon_{D0} \rightarrow \varepsilon_{DB0}$ $(W, Z)/G \rightarrow \tau, \nu_\tau$
Magical $\varepsilon_{D1}/\varepsilon_{D3}$ g/γ	Anger $\varepsilon_{D1}/\varepsilon_{D2}$ $g/(W, Z)$	Smell (Taste) ε_{D1} g	Sensuous Intuition $\varepsilon_{D1}/\varepsilon_{D0} \rightarrow \varepsilon_{DB0}$ $g/G \rightarrow t, b$
Archaic $\varepsilon_{D0}/\varepsilon_{D3}$ G/γ	Fear $\varepsilon_{D0}/\varepsilon_{D2}$ $G/(W, Z)$	Touch (Taste) $\varepsilon_{D0}/\varepsilon_{D1}$ G/g	Null Intuition ε_{D0} G

Table 11.1. Matrix of basic psychological functions, with dimensional values and particle correlates

The forms of thinking shown in the Table derive from the work of the cultural philosopher Jean Gebser (1985). Gebser described four basic types of cognition, each possessing its own dimensionality. The rational mode of thinking currently dominant in Western culture is three-dimensional; it is discursive, highly abstract, and is associated with empty space and linear time. Mythic thinking is a more concrete, earlier mode of functioning that Gebser regarded as two-dimensional. Characterized as oceanic, wavelike, or cyclical, mythic cognition flows in circular time, attuned to the rhythms of nature (waking-sleeping, the seasons, etc.) and continually harking back to the past. Still more primordial is one-dimensional magical thinking. In this timeless, largely unconscious state of mind, thinking operates in a purely associative way, by forming "sympathetic" linkages between elements based on their undifferentiated likenesses (Gebser 1985, p. 50). Zero-dimensional archaic cognition is most primal of all and is entirely embryonic, entailing but the bare possibility of thinking. Note importantly that *Topologies* departs from Gebser's interpretation of his lower dimensions as constituting independent dimensions in their own right. Instead the lower dimensions of thinking are taken as but immature or "fractional" forms of three-dimensionality (ε_{D3}), as "undertones" of it. And in the present book we see that each dimensional undertone has its boson ratio counterpart (cf. Tables 7.1 and 10.1).

The dimensional "eigenfunctions" of Table 11.1 appear on the principal diagonal of the matrix (as do the eigen-terms of our other matrices). It is with the emergence of these functions that the given forms of subjectivity reach maturity. In our musical metaphor, these self-functions constitute the "fundamental tones" of our dimensional symphony. Whereas the thinking function gains its maturity in its rational operations (ε_{D3}), feeling comes into its own through *love* (ε_{D2}). In *Topologies*, I make the case that love, anger, and fear are basic sub-dimensions of the two-dimensional lifeworld, with anger and fear being dimensional "undertones" of love. The "overtone" of love is termed *mental emotion*.

Mental emotion is a hybrid function that combines feeling with thinking. As such, it is the most abstract form of emotion. For example, when I delight over solving a difficult conceptual problem or become defensive when my views are challenged, mental emotion is at play. How does this type of emotion figure in the broader story of dimensional development? Whereas fear and anger are the psychic aspects of the immature, undertone forms of two-dimensionality, mental emotion is the psychic aspect of the overtone form. That is, mental emotion is the order of subjectivity involved in the support given by the two-dimensional emotional carrier wave to the generation of cognitive three-dimensionality. We know from Chapter 7 that, in making the transition from the third to fourth stage of projection, the development of the three-dimensional wave is aided by the introjection of the two-dimensional wave's $\varepsilon_{D3}/\varepsilon_{D2}$ overtone, which leads to the distillation of the ε_{DB2} bounding element for ε_{D3}. In Table 11.1, this distillation is signified by $\varepsilon_{D3}/\varepsilon_{D2} \rightarrow \varepsilon_{DB2}$; expressed in terms of particles, the transformation is from the $\gamma/(W, Z)$ boson ratio to the (e, ν_e) pair of fermions. What we are now seeing is the *psychical* aspect of this psychophysical operation. The subjective counterpart of the distilled fermion pair is mental emotion.

Does it still seem strange to relate physical events transpiring at the origin of the universe to familiar psychological functions? If so, let us not forget that — in a psychophysical universe — there must be psychological functions that are just as fundamental as the physical ones. While our everyday experiences indeed make many of the psychological functions given in Table 11.1 familiar to us, in their archetypal roots these functions are concrete universals constituting integral aspects of the "psychophysical atoms" of which our universe is composed. In the "atomic" substratum, the functions in question would not assume readily recognizable forms. An example from *Topologies* should prove instructive. There I noted that the familiar function of smell is rooted in a primal form of olfaction known as "chemoperception" (Boller 1995) that requires no specialized sense organ and operates through short-range chemical interactions occurring at the cellular level. I now hypothesize an even more primordial mode of olfaction operating at the sub-atomic level and associated with the gluonic interactions of the strong nuclear force (g). In a similar way, all the commonly experienced psychological functions of the Table should have their less familiar archetypal underpinnings.

Smell is in fact the eigenfunction of the one-dimensional lifeworld (ε_{D1}) underlying sensory experience. Touch is its dimensional undertone, and hearing and vision are dimensional overtones, functions that serve the one-dimensional vortex's selfless support of two- and three-dimensional cyclogenesis, respectively. (In *Topologies*, I demonstrate that taste is not an autonomous function in itself but takes a form associated either with smell or with touch.) Distillations of dimensional overtones into bounding elements, and corresponding boson ratios into fermion pairs, are again indicated by right-arrowed formulas. The hearing function is at play in the distillation of the $\varepsilon_{D2}/\varepsilon_{D1}$ overtone into the ε_{DB1} bounding element, which corresponds to the particle transformation of (W, Z)/g into (μ, v_μ). Vision is the psychic process involved in the distillation of the $\varepsilon_{D3}/\varepsilon_{D1}$ overtone into ε_{DB1}, or, in particle terms, γ/g into (u, d).

Finally, the "eigenfunction" of the zero-dimensional intuitive family is *null intuition* (ε_{D0}). The negative appellation reflects the fact that zero-dimensionality does not actually operate as an eigen- or self-function, since it *has no self*. Zero-dimensional intuition serves strictly in the capacity of a selfless carrier wave that facilitates higher-dimensional cyclogenesis through its overtones. Commonly speaking, sensuous intuition can be said to involve a deep bodily awareness that enables one to sense danger, for example, when no sensory cues are present (Vaughan calls this "physical intuition"; 1979, p. 66). Emotional intuition allows one to pick up on the overall feeling tone of a situation. The most abstract overtone of intuition is the mental variety. The classical intuition discussed in previous chapters is of this kind, and is exemplified in Descartes's definition of intuition: it is "an indubitable conception formed by an unclouded mind... [one that] is more certain even than deduction, because it is simpler....Thus, anybody can see by mental intuition that he himself exists, that he thinks, that a triangle is bounded by just three lines, and a globe by a single surface, and so on" (Descartes, 1628/1954, p. 155). It is the classical intuition of object-in-space-before-subject that has been proprioceptively reversed by the phenomenological intuition we have employed in this book. Topo-dimensionally understood, the several intuitive functions surely do not just reflect the experiences of particular individuals but the generic subjectivity of whole dimensions of cyclonic action. Each of these functions is a dimensional hybrid coupling intuition with a higher-dimensional function so as to promote the latter's Individuation. As before, transformations of dimensional overtones into bounding elements, and concomitant boson ratios into fermions, are given by right-arrowed formulae in Table 11.1.

Now, recall how the matrix of Table 5.2 was "set in motion" in Table 7.1 to provide a developmental account of dimensional process. Let us do the same for the matrix of Table 11.1.

By bringing out the psychical correlates of the physical particles given in Table 10.1, Table 11.2 provides a psychophysical summary of cosmic evolution that animates Table 11.1. The new table can be read in the same way as Table 10.1, since adding the subjective aspect does not change its developmental design. I will take note of just one novel feature. The psychical functions associated with overtone boson ratios and enclosed in curly brackets are precursory, incipient forms of the functions that are distilled with the fermion pairs.

Stage 4 (γ) Present Era $t > 10^{-12}$ sec.

Stage 5 (γ)

			γ rational thinking $[e, \nu_e]$ mental emotion

Stage 3 (W, Z; γ) Electroweak Era 10^{-36} to 10^{-12} sec.

Stage 6 (γ)　**Stage 4** (W, Z)

		W, Z love $[\mu, \nu_\mu]$ hearing	(W, Z)/γ mythic thinking [u, d] vision
		γ/(W, Z) {mental emotion}	

Stage 2 (g; W, Z; γ) GUT Era 10^{-43} to 10^{-36} sec.

Stage 7 (γ)　**Stage 5** (W, Z)　**Stage 3** (g)

	g smell [t, b] sensuous intuit.	g/(W, Z) anger $[\tau, \nu_\tau]$ emot. intuition	g/γ magical thinking [c, s] ment. intuition
	(W, Z)/g {hearing}		
	γ/g {vision}		

Stage 1 (g; W, Z; γ) Planck Era $t < 10^{-43}$ sec.

Stage 8 (γ)　**Stage 6** (W, Z)　**Stage 4** (g)

G Null intuition	G/g touch	G/(W, Z) fear	G/γ archaic thinking
g/G {sensuous intuition}			
(W, Z)/G {emotional intuition}			
γ/G {mental intuition}			

Table 11.2. The psychophysics of cosmogony

11.2 Toward a Reflexive Physics

A wide variety of approaches have been taken to the problems of force unification and cosmogony that confront contemporary physics. Yet, whatever their differences, these theoretical initiatives have all been tacitly guided by the long-standing, deeply engrained presupposition of object-in-space-before-subject. In the posture that physicists have always assumed, force fields and particles — and even space and time themselves — are regarded as objective entities implicitly embedded in a more abstract analytical space, with the analyst him- or herself being detached from the entities analyzed. The ideas I have set forth in this book suggest that a basically different posture is required if we are to effectively address the most pressing issues of modern physics. I would now like to further explore the question of what the new posture entails. To prepare for this, let us focus on a characteristic common to physics and phenomenology alike: inherent reflexivity.

Whereas science's conventional posture presupposes the division of subject and object, I have been emphasizing the sub-objective or psychophysical character of topodimensional action. Here subjective and objective aspects of dimensional action interpenetrate one another dynamically, and this makes for action that is self-referential or reflexive. Psychical and physical aspects are linked so intimately that the movement from one to the other is at once a return to the first. In the words of Merleau-Ponty, such dialectical action implies that "each term is itself only by proceeding toward the opposed term, becomes what it is through the movement, that it is one and the same thing for each to pass into the other or to become itself, to leave itself or to retire into itself" (1968, pp. 90–91). This reflexive circulation of subject and object is graphically illustrated in Figure 4.2, Ryan's diagram of the Klein bottle, which I have adapted to show the unbroken flow from uncontained subject to contained object to containing space. Each of the sixteen psychophysical atoms of dimensional action considered in the previous section possesses this reflexive quality.

The fundamental role of reflexivity was implied earlier in our finding that the microphysical action of the quantum realm and the depth-dimensional action described by Merleau-Ponty are tied to each other via the prereflective psychophysical action of self-referential Kleinian spin (see Chapter 4). That reflexivity also plays a key role in cosmogony is of course indicated by the title of this book: *The Self-Evolving Cosmos*. Cosmodimensional organisms evolve autopoietically, by giving birth to themselves. In the process of dimensional Individuation, action spheres first diverge, their reflexive nature being obscured. This is followed by the convergence of dimensional spheres wherein braneworlds complete their Individuation in relation to one another by returning to themselves in proprioceptive self-awareness.

Recognizing the thoroughgoing reflexivity of the phenomena of contemporary physics brings a clearer understanding of what underlies the difficulties physicists have encountered in their attempts at unification. For, they have approached their phenomena in a non-reflexive way, seeking to objectify processes that are fundamentally resistant to such an effort. It would not be enough for physicists merely to acknowledge the subjective side of their phenomena while tacitly maintaining classical science's stance of analytical detachment. Since physics and cosmogony entail primordial actions in which subject and object inseparably fuse, it seems futile for the analyst of these processes to continue in the posture of a detached subject before whom particular objects are cast. In approaching the phenomena on their own terms, the analyst evidently must enter into them with his or her subjectivity. No longer can s/he remain a disinterested bystander, for her active presence is required to complete the analysis in a concrete way.

Perhaps it is not too much to say that the present book — through its use of phenomenological intuition and topological imagination — makes a preliminary contribution toward a fully reflexive physics. To gain a better understanding of what such a physics would involve, let us turn to the work of the philosopher of science Evelyn Fox-Keller.

Fox-Keller calls for a new form of perception in scientific inquiry that she names "dynamic objectivity" (1985, p. 115). The old approach, she says, involves a "static objectivity" in which "the pursuit of knowledge... begins with the severance of subject from object" (p. 117). In contrast,

> dynamic objectivity aims at a form of knowledge that grants to the world around us its independent integrity but does so in a way that remains cognizant of, indeed relies on, our connectivity with that world. In this, dynamic objectivity is not unlike empathy, a form of

knowledge of other persons that draws explicitly on the commonality of feelings and experience in order to enrich one's understanding of another in his or her own right. (1985, p. 117)

Dynamic objectivity employs a type of awareness closely akin to what I have called proprioception. This is further evidenced in Fox-Keller's citation of Piaget: "'Objectivity consists in...fully realizing the countless intrusions of the self in everyday thought and the countless illusions which result....So long as thought has not become conscious of self, it is a prey to perpetual confusions between objective and subjective'" (p. 117). According to Fox-Keller:

Dynamic objectivity is thus a pursuit of knowledge that makes use of subjective experience (Piaget calls it consciousness of self) in the interests of a more effective objectivity. Premised on continuity [of self and other], it recognizes difference between self and other as an opportunity for a deeper and more articulated kinship. The struggle to disentangle self from other is itself a source of insight — potentially into the nature of both self and other....To this end, the scientist employs a form of attention to the natural world that is like one's ideal attention to the human world: it is a form of love. The capacity for such attention, like the capacity for love and empathy, requires a sense of self secure enough to tolerate both difference and continuity. (1985, pp. 117–18)

Writing in the same vein, Fox-Keller adduces Ernest Schachtel's distinction between "autocentric" and "allocentric" perception. Whereas the former is "dominated by need or self-interest," the latter "is perception in the service of a love 'which wants to affirm others in their total and unique being.' It is an affirmation of objects as 'part of the same world of which man is a part,'" one which "permits a fuller, more 'global' understanding of the object in its own right" (p. 119). Although Fox-Keller pays scant attention to phenomenological philosophy, citing none of its leading figures, the main thrust of her presentation is much in keeping with phenomenology's central aim, as expressed in its well-known slogan: "To the things themselves!"

And it seems clear that the world shared by the "allocentric" observer and the objects that s/he observes is the *lifeworld of phenomenology*.

Fox-Keller helps us gain a better grasp of the new mode of scientific inquiry by offering a specific example of one of its premier practitioners: the Nobel prize-winning biologist, Barbara McClintock. In stark contrast to the detached, dispassionate attitude of the Cartesian scientist, McClintock speaks of obtaining an intimate feeling for the plants she works with: "'I don't feel I really know the story if I don't watch the plant all the way along. So I know every plant in the field. I know them intimately, and I find it a great pleasure to know them.'" (Fox-Keller 1985, p. 164). In another place, McClintock

> describes the state of mind accompanying the crucial shift in orientation that enabled her to identify chromosomes she had earlier not been able to distinguish: "I found that the more I worked with them, the bigger and bigger [the chromosomes] got, and when I was really working with them I wasn't outside, I was down there. I was part of the system....It surprised me because I actually felt as if I was right down there and these were my friends....As you look at these things, they become part of you. And you forget yourself." (McClintock quoted in Fox-Keller 1985, p. 165)

Fox-Keller observes that McClintock's vocabulary "is consistently a vocabulary of affection, of kinship, of empathy," an empathy that constitutes "the highest form of love: love that allows for intimacy without the annihilation of difference" (p. 164). Here the word "love" is used "neither loosely nor sentimentally, but out of fidelity to the language McClintock herself uses to describe a form of attention, indeed a form of thought" (p. 164).

Fox-Keller arrives at these conclusions:

> The crucial point for us is that McClintock can risk the suspension of boundaries between subject and object without jeopardy to science precisely because, to her, science is not premised on that division. Indeed, the intimacy she experiences with the objects she studies...is a wellspring of her powers as a scientist....In this world of difference, division is relinquished without generating chaos. Self and other, mind

and nature survive not in mutual alienation, or in symbiotic fusion, but in structural integrity. (1985, pp. 164–65)

Finally, after recounting the goal of conventional science, Fox-Keller observes that, "To McClintock, science has a different goal: not prediction per se, but understanding; not the power to manipulate, but empowerment — the kind of power that results from an understanding of the world around us, that simultaneously reflects and affirms our connection to that world" (p. 166).

In phenomenological terms, the world to which McClintock is connected in feeling and embodied empathy is the lifeworld. It is a world in which the dialectic of difference and identity is enacted proprioceptively, through an intimate knowledge of other that requires and is inseparable from the knowledge of self (a "consciousness of self"). McClintock's "revolution that 'will reorganize...the way we do [scientific] research'" (Fox-Keller 1985, p. 172) depends upon descending from the Cartesian stratosphere, relaxing our commitment to the classical ideal of the continuum, of pure positivity, "pure light." We immerse ourselves now in the nether dimension of *depth*, where light and shadow mingle, where continuity and discontinuity, object and subject, mediate one another internally in an encompassing circular flow.

Fox-Keller's "dynamic objectivity" as exemplified by McClintock is hardly the only instance of the burgeoning of a new dialectical science. The phenomenological initiative, begun early in the twentieth century, has been advanced by thinkers like Heelan (1983) and Gendlin (1991), who have proposed that the work of science not proceed from "stratospheric" perception, but from the intricacies of the lifeworld or lived body. As we saw in Chapter 3, a dialectical approach to science also is advocated by Tanabe Hajime (1986), whose own version of proprioception is termed "metanoetics." More recently, biophysicist Koichiro Matsuno (1995) has called for a "dialogical" science that would supersede the old "monologue" carried on by the solitary Cartesian subject who looks down upon the world from above. In Matsuno's vision, scientific activity would involve a *community* of subjects concretely engaged with each other in dynamic and generative negotiations. Whereas the Cartesian subject is anonymous, absent from the events that transpire, the participants in the dialectical community would function self-referentially (proprioceptively) to include themselves in the process (Matsuno exemplifies this by explicitly including himself as author in what he writes; 1995, 1998). A new contribution to emergent dialectical science is offered by the Jungian psychologist Nathan Schwartz-Salant (2007). Operating self-referentially, Schwartz-Salant employs Merleau-Ponty and the Klein bottle in characterizing the deep psychodynamics of human relationships, and he likens the fields operative in these paradoxical interactions to field processes in fundamental physics.

What we require in the present context is a dialectical *cosmogony*. Part of this involves situating the analysis of cosmogony within cosmogony itself. The unquestioning objective stance analysts have tended to take toward cosmic evolution is in fact a product of a certain stage of evolution, namely, the fourth and final stage of projection. This is the stage of psychophysical development whose psychical aspect is governed by an abstract and objectifying mode of thought (see Table 11.2). It is here that we assume that cosmogonic events are "objectively out there," and that we analysts are detached from them, with our lived subjectivity playing no role. In this stage, the common sense notion of a "universe out there" developing on its own is so compelling that it seems absurd for us to think otherwise. But, in advancing to the stages of proprioception, the point comes home to us that we are indeed intimate participants in the story of cosmic Individuation. Thus entering into cosmogonic process, the classical posture of analysis gives way to a phenomenological one in which our *own* process of development plays an integral role. In the act of inwardly grasping the transformation of the cosmos, the analyst surpasses the projective construction of herself as an isolated onlooker and takes part in the drama of creating a world. So, if the cosmos is self-evolving, the self of the *analyst* figures essentially in the reflexive enactment of this process.

But let me try to be clearer about what the involvement of the analyst specifically entails. The proposition I offer is that a fully reflexive analysis of cosmogony requires that, in investigating the stages of cosmogonic proprioception, the analyst must proprioceive *his or her own stages of development*. Only then can the link to cosmic development be realized in its existential immediacy, since, only then would the analyst realize cosmic Individuation as a *self*-Individuation. Of course, the analytical self in question cannot merely be that of a finite particular individual. The self that participates in the concrete universality of cosmic transformation must also be universal. Yet it seems we need to begin with the particular person if the process is to be grounded in existential reality. Presumably, in the course of deeply exploring his or her own past, the analyst would cross a threshold and her personal being would shade into the transpersonal. The transpersonal psychiatrist Stanislav Grof expressed a similar idea in describing the transformation of awareness that can occur in the act of re-experiencing the "perinatal" stages of development, those occurring around the time of birth: "All we can say is that somewhere in the process of confrontation with the perinatal level of the psyche, a strange qualitative Moebius-like [!] shift seems to occur in which deep self-exploration of the individual unconscious turns into a process of experiential adventures in the universe-at-large" (1985, p. 36).

In the projective moment of cosmogony, it may well seem a flight of fancy to link the stages of human development to those of the cosmos as a whole. The proprioceptive response to this incredulity extends the biological dictum that "ontogeny recapitulates phylogeny" to the field of physics and says, ontogeny recapitulates *cosmogony*. For, if it is true that we participate in the story of creation in a full-fledged way, it would seem that our own history would be inseparable not only from that of the broader biological world but from nature at large. Evidently then, to gain a proprioceptive grasp of the fields and particles of nature's archaic past, to apprehend them in the most concrete, immediate, and deeply reflexive way, it seems the analyst must work through his or her own archaic past.

Embryological research certainly appears to support the idea that the early development of the human individual mirrors the development of the species as a whole. In fact, the work done in *Topologies of the Flesh* links ontogeny and phylogeny explicitly, and in a detailed way. What the present work calls for is a linking of ontogeny and cosmogony. Some theorists have broadly speculated that the universe functions as a giant hologram (Bohm 1980, p. 189). Such a cosmos should possess a fractal pattern of self-similarity, with the structure and development of the whole being mirrored recursively on every scale of magnitude down to the smallest part. Then — if probing the early history of an individual member of the phylogenetic order opens out into phylogeny as a whole — it is perhaps not unreasonable to hypothesize a deeper stratum of self-similarity involving the history of the cosmos to which we belong. Relevant in this regard is the vision of physicist Lee Smolin:

> Living things share in some ways, and extend in other ways, the basic properties of non-equilibrium self-organized systems that seem to characterize the universe on every scale, from the cosmos as a whole to the surface of planets....If life, order and structure are the natural state of the cosmos itself, then our existence, indeed our spirit, might finally be comprehended as created naturally, by the world, rather than unnaturally and in opposition to it. (1997, p. 160)

In a similar vein, biophysical theorist Hector Sabelli asserts that "the continuity of evolution requires that the same fundamental forms must be expressed at the physical, biological, and psychological levels of organization" (2005, p. 431). This is consistent, of course, with the psychophysical nature of cosmogonic process. Having spelled out the correlations between psyche and physis in the preceding section, we are now prepared to consider the specific means of forging the link between microcosm and macrocosm — between the analyst's development and that of the cosmos.

11.3 Concretization of the Self-Evolving Cosmos

We know that full-fledged participation in cosmogony means realizing cosmic Individuation as self-Individuation. To bring this about, the analyst must enter into each Individuating psychophysical quantum of action via a reflexive act of proprioception. In so doing, the proprioceptions of the dimensional organisms described in previous chapters are made into concrete realities. A process of this kind was adumbrated in *Topologies of the Flesh*. Dimensional lifeworlds gained more tangible expression through the author's own reflexive activities. Of course, *Topologies* did not deal explicitly with particles, fields, and cosmogony. In what follows, the proprioceptions specified in *Topologies* will be adaptively reprised so as to provide an indication of how the self-evolving cosmos — its harmonies, its unity-in-diversity — can be realized more concretely.

*

In the classical posture of scientific detachment, all analysis is analysis of what is other. What we require now, however, is an analysis of self. The first step the analyst must take toward this end is descending from the conceptual stratosphere, lifting his veil of anonymity, and standing present in his or her embodied actuality. Without such a concretization — where the analyst makes his presence real by situating himself within his own analysis — there can be no genuine *self*-analysis. Let me acknowledge then, that, behind all the abstract ideas presented in this book, there is Steven Rosen, a concrete person whose existential being is grounded in his body. It is through this body of mine that the requisite self-analysis must be routed. In the course of bringing my awareness backward into my finite particular body, in properly proprioceiving that body, the self-analysis eventually opens out into the concrete universality of the *dimensionally* embodied selves, the generic organisms composed of sixteen psychophysical quanta of action.

The initial proprioceptions involve thinking and electromagnetism, mental emotion and the electron–electron neutrino fermion pair (see Table 11.2). How is the body to be entered for this? Through the head, I propose.

It is a commonplace of neuroscience that cognitive activity is linked to the cortical region of the brain. In *Topologies*, I drew on the work of psychiatrist Trigant Burrow to demonstrate that it is possible to *proprioceive* the workings of the brain. Supplementing his laboratory research with systematic self-observation, Burrow kinesthetically detected a unique pattern of tensions around the eyes and forehead that he found to be correlated with the symbolic operations of the cerebral cortex. In this experiential self-probing of the brain's "symbolic segment" (1953, p. 316), the area that constitutes the site of thinking and language, Burrow was exploring the seat of his own identity as an individual governed by the thinking function. Or, in Merleau-Ponty's terms, Burrow was investigating the "*I think* that must be able to accompany all our experiences" (Merleau-Ponty 1968, p. 145). Burrow himself came to refer to this cortical center of identity as the *"I"-persona*. This thinking subject that presides over every facet of human experience and behavior should not be confused with the ego of the allegedly isolated individual. We might say that the "I"-persona is the *species-wide* "subject" that lies behind the appearance of individual subjectivity — the subjectivity of "Steven Rosen," for example. But while it is through the "I"-persona that we, as a species, create the impression of ourselves as merely isolated, disembodied subjects, the generic "I" itself is no disembodied subject. Rather, it is the *bodily process* that is central to human functioning as a whole. Therefore, when Burrow became attentive to the "I"-persona rather than continuing to be unwittingly governed by it, he experienced this palpable pattern of tension around the eyes and forehead against the "tensional pattern of the organism as a whole" (Galt 1995, p. 31). He was thus presumably able to apprehend in an immediate way what he called the "phyloörganism" (p. 445), i.e., the organism of humanity at large. Burrow coined the term "cotention" (1932) for his practice of proprioceiving the generic organism. He described the procedure as one of setting aside daily experimental periods in which he "adhered consistently to relaxing the eyes and to getting the kinesthetic 'feel' of the tensions in and about the eyes and in the cephalic area generally" (1953, p. 95).

Proceeding in the manner of Burrow, I, the psychophysical analyst, am to enter my body through my head to obtain a bodily sense of the head that is directing this analysis. Since the proprioception thereby enacted is no act of "pure meditation" that leaves thinking and language behind, it seems the process must include the appropriate mode of signification if the maximum effect is to be achieved. In the semiosis required, neither words nor conventional mathematical symbols will suffice. What is needed is that unique topological signifier that refers concretely to the "fourth dimension" — the dimension of depth incorporating my subjectivity — by referring to *itself*. I am speaking, of course, of the Klein bottle. Assuming the Klein bottle is not just taken as a signified topological object or as an arbitrarily devised, conventionally agreed upon signifier (as are most mathematical symbols), it is this body of paradox that constitutes the semiotic content of the proprioceived brain. In the proprioception of the brain that is at once a phenomenological meditation upon the Klein bottle, the analyst surpasses the brain of the particular individual to gain a glimmer of the dimensional organism's "generic brain," the "braneworld" in which the thinking function is centered. The brain thus proprioceived is no mere object of scientific scrutiny but is the sub-objectively *lived* brain (Leder 1990, p. 113); it is the brain as a concrete universal, as the "flesh of the world" (Merleau-Ponty 1968, p. 139).

In its physical aspect, the proprioception of the brain involves electromagnetism or light. At the beginning of this chapter, I paraphrased Heidegger in noting the close relationship between thinking and light: "to think Being is to think light." This suggests that, in thinking light, we engage in an ultimate *self*-thinking. In Chapter 3, and again in Chapter 10, I noted that the light cones of Einstein-Minkowski space-time stretch out from a subjective here-now origin, the perspective point of the observer. This subject at the center of the light cones is our *thinking* subject. In the earlier chapters, I indicated that, in making the transition from Einsteinian space-time to the phenomenological lifeworld, we must switch gears and move backward into the subjective source of the light cones, which would amount to retracting the light cones, reeling them back into their here-now origin. That is to say, light must be proprioceived.

The initial set of proprioceptions indicated in Table 11.2 does call for more than the proprioception of light or electromagnetism (γ). We have found that electromagnetic action is the global aspect of a three-dimensional Kleinian braneworld whose local aspect consists of the electron–electron neutrino pair of dimensional bounding elements (e, ν_e). In order to transform this fermionic action into *self*-action, I must include my own subjectivity in it, which means proprioceiving the psychic correlate of said action, namely, mental emotion.

According to Burrow, the cerebral cortex of the brain that houses the thinking function is also the seat of the "affecto-symbolic" mode of functioning (1953, pp. 530–31). This abstract form of emotional expression closely related to thinking is what we are presently calling mental emotion. In preparing for the proprioception of said emotion, I acknowledge the motivational basis of this analysis.

With this book, I hope to contribute to our understanding of the universe and its psychophysical evolution. Yet behind my lofty aspiration there is also a strong desire to win the recognition of my colleagues, thereby reinforcing my self-image. So while there are moments when I might like to think myself capable of proceeding in a purely dispassionate way that is detached from my "baser" feelings and self-centeredness, the fact is that every word I have written, however rarefied and cerebral, and these very words I am now writing, are colored by an emotional, self-possessed undertone.[2] Does this recognition of the mental emotion that undergirds the present analysis constitute a full-fledged proprioception of that emotion? Clearly it does not. To engage in said proprioception, it will not suffice to take back the particular self-oriented emotions I experience. Beyond that, the "I" itself must be taken back. It is in topodimensionally withdrawing the projection of "Steven Rosen" as the prime container of mental emotion that the emotions accompanying "my" analysis in fact become the experiences of humanity at large — that is, of the Kleinian lifeworld. Apparently then, "I" must obtain a palpable internal impression of the region of "my" head from which originates "my" present feeling of frustration at the difficulty of conveying these ideas — and the subsequent feeling of self-satisfaction when success seems achieved. Proprioceptions of this kind, insofar as they lead backward (through "Steven Rosen") to the "I"-persona as such, open a pathway to the concrete universal.

Actually, in stage 5 of electromagnetic development, there is another type of proprioception not directly indicated in Table 11.2. We know from the previous chapter that as development progresses through the stages of projection, the dimensional actions of earlier stages are not merely left behind but persist in the form of *penumbras*. Consider the transition from the third to fourth stage of projection. Although mythic thinking and visual perception are both carried forward as penumbras, the former is an undeveloped function that continues as but an undertone of rational thinking, whereas the latter, being an overtone function, advances in a *sublimated* form: while the visual experience of the third stage of projection is diffuse, dreamlike, and imaginal, by the fourth stage it has become the sharply focused, abstract kind of visualization occurring in Cartesian space (see Rosen 2006, p. 254). Evidently this has consequences for the process of proprioception. Stage 6 of electromagnetic development entails a proprioception of the original, unsublimated form of visual perception. Before this can happen, it appears that a preliminary proprioception of the sublimated form of vision must take place in stage 5. In a similar way, the proprioception of primary mental intuition in stage 7 must be preceded by two preliminary proprioceptions of sublimated forms of mental intuition. The first of these occurs in electromagnetic stage 5 and in fact entails the *phenomenological* intuition we have applied in previous chapters. This has been our means of reversing the classical intuition of object-in-space-before-subject, which we now recognize as a highly sublimated form of

mental intuition. In stage 6, a preliminary proprioception of a less sublimated form of mental intuition is required.

In Chapter 8 of *Topologies*, I provided a detailed psycho-dimensional
account of all four rounds of proprioception. Having reprised a portion of
this here, I will presently limit myself to merely summarizing the remainder of the work done in that chapter. However, the summary will include
the physical aspects of the proprioceptions and will place the analysis in
the context of cosmogony.

The second full set of proprioceptions is that associated with stage 6 in
the genesis of electromagnetism, and stage 4 in the development of the
weak force. With respect to the former, the fractional, $(W, Z)/\gamma$ form of
electromagnetism is proprioceived, along with its up and down quark
bounding elements. The psychic aspect of this challenges the analyst to
consciously engage an older, hitherto repressed facet of his or her subjectivity, that involving the old mammalian brain and the mythic thinking and
imaginal vision of the child (in the concrete actualization carried out in
Topologies, my own "inner child" made his presence felt in the person of
"Stevie"). This proprioception of the immature three-dimensional lifeworld (which includes a preliminary proprioception of mental intuition) is
coupled with a backward movement into the two-dimensional lifeworld
that involves the weak force and the function of love, accompanied by the
μ and v_μ bounding elements, whose psychic aspect is hearing. In the joint
proprioceptions carried out, the universe draws back into itself, contracts,
as Kleinian and Moebial braneworlds come into harmony.

The third round of proprioceptions brings us to stage 7 of electromagnetic Individuation, stage 5 in the genesis of the weak force, and stage 3 in
the development of the strong force. The electromagnetic proprioception
is of g/γ and its charm and strange quark bounding elements. In psychic
terms, the analyst is called on to lift the repression of a still older aspect of
his or her subjectivity, one that entails the reptilian brain, magical thinking, and primary mental intuition. Together with this proprioception of the
nascent Kleinian braneworld, two lower-dimensional proprioceptions take
place. In the two-dimensional Moebial world, the fractional weak force
$(g/(W, Z))$ correlated with the emotion of anger is proprioceived, as is the
τ–v_τ pair of bounding elements linked to emotional intuition. In the one-
dimensional lemniscatory world, the strong force (g) connected psychi-

cally to smell is taken back in, along with its top and bottom quark bounding elements, which are related to sensuous intuition. This tripartite round of proprioceptions attests to the deepening of cosmic contractions and enhancement of dimensional harmony, as the symphony of the spheres is now in full swing.

In the final round of backward circulations, the higher-dimensional branewaves complete their self-evolution in relation to each other, and to the selfless zero-dimensional carrier wave that hitherto had been repressed. Stage 8 of electromagnetic Individuation brings proprioception of the primal state of affairs wherein the embryonic photon, G/γ, rests in the womb of gravitation, supported by the γ/G gravitational overtone. Here the analyst faces the profound challenge of gaining conscious access to the deepest recesses of his or her subjectivity, those involving the primitive brain stem (neural chassis) and the archaic form of thinking. In conjunction with the proprioception of the incipient Kleinian branewave, two lower-dimensional proprioceptions are enacted. Reflexive awareness is gained of the fractional Moebial wave $(G/(W, Z))$ linked to fear, and of the embryonic lemniscatory wave (G/g), whose psychic aspect is touch. No proprioceptive self-realization is associated with the sub-lemniscatory wave as such. The zero-dimensional gravitational wave (G) is realized only in its capacity as the selfless carrier that supports the self-development of the higher-dimensional waves.

With the fourth round of proprioceptions, cosmic contractions are complete. Dimensional organisms have drawn back in upon themselves in concert with one another. Having moved fully backward into the "black holes" constituting their own "birth canals," they have harmonically converged. And through the synchronous action of the reverse self-nativities, the embryo of a new dimensional organism is formed to signal the opening of a wider turning in nature's spiral of creative growth.

Notes

1. In *Topologies*, the term "eigenfunction" is used in a qualitative sense distinctly different from its quantitative usage in the mathematical sciences.

2. Contemporary brain research and phenomenological investigation alike indicate that cognitive activity cannot be treated in simple separation from emotions without risking an oversimplification of human behavior and experience (Ellis 1997, Suvin 1997).

Bibliography

Abbott, Edwin. 1884/1983. *Flatland: A Romance of Many Dimensions*. New York: Barnes and Noble.

Abram, David. 1996. *The Spell of the Sensuous*. New York: Vintage.

Angeles, Peter A. 1981. *Dictionary of Philosophy*. New York: Barnes and Noble.

Applebaum, David. 2000. "Dirac Operators — From Differential Calculus to the Index Theorem." Based on inaugural lecture as Professor of Mathematics at the Nottingham Trent University. http://www.facct.ntu.ac.uk/staff/personal/dapplebaum/absconc.pdf (accessed May 8, 2004; link currently dead).

Babich, Babette E., ed. 2004. *Hermeneutic Philosophy of Science, Van Gogh's Eyes, and God*. Dordrecht: Kluwer.

Barr, Stephen. 1964. *Experiments in Topology*. New York: Dover.

Bigwood, Carol. 1993. *Earth Muse*. Philadelphia: Temple University Press.

Bohm, David. 1965. *The Special Theory of Relativity* (Appendix). New York: Benjamin.

———. 1980. *Wholeness and the Implicate Order*. London: Routledge and Kegan Paul.

——— . 1994. "The Bohm/Rosen Correspondence." In *Science, Paradox, and the Moebius Principle*, edited by Steven M. Rosen, 223–58. Albany: State University of New York Press.

Bohm, David, and Basil J. Hiley. 1975. "On the Intuitive Understanding of Non-Locality as Implied in Quantum Theory." *Foundations of Physics* 5:93–109.

——— . 1984. "Generalization of the Twistor to Clifford Algebras as a Basis for Geometry." *Revista Brasiliera de Fisica* July:1–26.

Boller, Thomas. 1995. "Chemoperception of Microbial Signals in Plant Cells." *Annual Review of Plant Physiology and Plant Molecular Biology* 46:189–214.

Burrow, Trigant. 1932. *The Structure of Insanity*. London: Kegan Paul, Trench, Trubner and Co.

——— . 1953. *Science and Man's Behavior*. New York: Philosophical Library.

Candelas, Philip, Gary T. Horowitz, Andrew Strominger, and Edward Witten. 1985. "Vacuum Configurations for Superstrings." *Nuclear Physics* B258:46–74.

Čapek, Milič. 1961. *Philosophical Impact of Contemporary Physics*. New York: Van Nostrand.

Carter, Brandon. 1968. "Global Structure of the Kerr Family of Gravitational Fields." *Physical Review* 174:1559–71.

Cataldi, Susan L. 1993. *Emotion, Depth, and Flesh: A Study of Sensitive Space*. Albany: State University of New York Press.

Crease, Robert. 1997. "Hermeneutics and the Natural Sciences: Introduction." *Man and World* 30:259–70.

Cremmer, Eugene and Joël Scherk. 1976. "Dual Models in Four Dimensions with Internal Symmetries." *Nuclear Physics* B103:399–425.

Descartes, René. 1628/1954. "Rules for the Direction of the Mind." In *Descartes's Philosophical Writings*, edited and translated by Elizabeth Anscombe and P. T. Geach, 153–80. London: Thomas Nelson and Sons.

Eddington, Arthur Stanley. 1946. *Fundamental Theory*. Cambridge: Cambridge University Press.

Eger, Martin, 2006. *Science, Understanding, and Justice*. Chicago: Open Court.

Ellis, Ralph. 1997. "Differences between Conscious and Non-Conscious Processing." Background paper for After-Postmodernism Conference, University of Chicago, November 14–16. http://www.focusing.org/apm_papers/ellis.html.

Englert, François, Laurent Houart, and Anne Taormina. 2002. "The Bosonic Ancestor of Closed and Open Fermionic Strings." Cornell University online archive (arXiv): hep-th/0203098, v1, March 11.

Ernst, Bruno. 1986. *Der Zauberspiegel des M.C. Escher*. Berlin: Taco.

Fehér, Márta, Olga Kiss, and László Ropolyi, eds. 1999. *Hermeneutics and Science*. Dordrecht: Kluwer.

Foerster, Heinz von. 1976. "Objects: Tokens for (Eigen-)Behaviors." *ASC Cybernetics Forum* 8:91–96.

Fox-Keller, Evelyn. 1985. *Reflections on Gender and Science*. New Haven, CT: Yale University Press.

Frescura, F. A. M., and Basil J. Hiley. 1980. "The Implicate Order, Algebras, and the Spinor." *Foundations of Physics* 10:7–31.

Freud, Sigmund. 1914/1957. "On Narcissism: An Introduction." In *Standard Edition of the Complete Works of Sigmund Freud*, vol. 14, edited by James Strachey, 117–40. London: Hogarth Press.

Fuller, B. A. G. and Sterling M. McMurrin. 1957. *A History of Philosophy*. New York: Henry Holt and Co.

Gallagher, Shaun, and Jonathan Shear, eds. 2000. *Models of the Self*. Thorverton, UK: Imprint Academic.

Galt, Alfreda. 1995. "Trigant Burrow and the Laboratory of the 'I.'" *The Humanistic Psychologist* 23:19–39.

Gebser, Jean. 1985. *The Ever-Present Origin*. Athens, OH: Ohio University Press.

Gendlin, Eugene T. 1991. "Thinking Beyond Patterns: Body, Language, and Situations." In *The Presence of Feeling in Thought*, edited by Bernard denOuden and Marcia Moen, 27–189. New York: Peter Lang.

Gendlin, Eugene T., and Jay L. Lemke. 1983. "A Critique of Relativity and Localization." *International Journal of Mathematical Modelling* 4:61–72.

Glazebrook, Trish. 2000. *Heidegger's Philosophy of Science*. New York: Fordham University Press.

Graves, John C. 1971. *The Conceptual Foundations of Contemporary Relativity*. Cambridge, MA: MIT Press.

Greene, Brian. 1999. *The Elegant Universe*. New York: W. W. Norton.

———. 2004. *The Fabric of the Cosmos*. New York: Alfred A Knopf.

Gribbin, John. 2000. *Q is for Quantum*. New York: Simon and Schuster.

Grof, Stanislav. 1985. "Modern Consciousness Research and Human Survival." *Re-Vision* 8:27–39.

Grosz, Elizabeth. 1994. *Volatile Bodies*. Bloomington: Indiana University Press.

Gutting, Gary, ed. 2005. *Continental Philosophy of Science*. Oxford: Blackwell.

Heelan, Patrick A. 1983. *Space-Perception and the Philosophy of Science*. Berkeley: University of California Press.

———. 1997. "After Post-Modernism: The Scope of Hermeneutics in Natural Science." Background paper for After-Postmodernism Conference, University of Chicago, November 14–16. http://www.focusing.org/apm_papers/heelan.html.

Heidegger, Martin. 1927/1962. *Being and Time*, translated by John Macquarrie and Edward Robinson. New York: Harper and Row.

———. 1946/1984. *Early Greek Thinking*, translated by David F. Krell and Frank A. Capuzzi. New York: Harper and Row.

———. 1954/1971. "The Thinker as Poet." In *Poetry, Language, Thought*, translated by Albert Hofstadter, 1–14. New York: Harper and Row.

———. 1962/1972. "Time and Being." In *On Time and Being*, translated by Joan Stambaugh, 1–24. New York: Harper and Row.

———. 1964/1977. "The End of Philosophy and the Task of Thinking." In *Martin Heidegger: Basic Writings*, edited by David F. Krell, 373–92. New York: Harper and Row.

Helmholtz, Hermann. 1877/1954. *On the Sensations of Tone as a Physiological Basis for the Theory of Music*, translated by Alexander J. Ellis. New York: Dover.

Hesse, Mary. 1965. *Fields and Forces*. Totowa, NJ: Littlefield, Adams and Co.

Hestenes, David. 1983. "Quantum Mechanics from Self-Interaction." *Foundations of Physics* 15:63–87.

Hofstadter, Douglas. 1979. *Gödel, Escher, Bach: An Eternal Golden Braid*. New York: Basic Books.

Horava, Petr, and Edward Witten. 1996. "Heterotic and Type I String Dynamics from Eleven Dimensions." *Nuclear Physics* B460:506–24.

Hu, Huping, and Maoxin Wu. 2004. "Spin as Primordial Self-Referential Process Driving Quantum Mechanics, Spacetime Dynamics and Consciousness." *Neuroquantology* 2: 41–49.

Hurewicz, Witold, and Henry Wallman. 1941. *Dimension Theory*. Princeton, NJ: Princeton University Press.

Husserl, Edmund. 1936/1970. *Crisis of the European Sciences and Transcendental Phenomenology*, translated by David Carr. Evanston, IL: Northwestern University Press.

Iran-Nejad, Asghar. 1989. "A Nonconnectionist Schema Theory of Understanding Surprise-Ending Stories." *Discourse Processes* 12:127–48.

Jahn, Robert, and Brenda Dunne. 1984. *On the Quantum Mechanics of Consciousness* (Appendix B). Princeton, NJ: Princeton University School of Engineering/Applied Sciences.

Jeannerot, Rachel, Jonathan Rocher, and Mairi Sakellariadou. 2003. "How Generic is Cosmic String Formation in SUSY GUTs?" *Physical Review* D68 103514 (hep-ph/ 0308134).

Josephson, Brian D. 1987. "Physics and Spirituality: The Next Grand Unification?" *Physics Education* 22:15–19.

Jung, Carl Gustav. 1961. *Memories, Dreams, Reflections*, recorded and edited by Aniela Jaffé; translated by Richard and Clara Winston. New York: Vintage.

———. 1970. *Mysterium Coniunctionis*. Vol. 14 of *Collected Works*, translated by R. F. C. Hull. Princeton, NJ: Princeton University Press.

———. 1971. *Psychological Types*. Vol. 6 of *Collected Works*, translated by R. F. C. Hull. Princeton, NJ: Princeton University Press.

Kaluza, Theodor. 1921. "Zum Unitätsproblem der Physik." *Sitzungsberichte der Preussische Akademie der Wissenschaften* LIV:966–72.

Kiehn, R. M. 1999. "An Extension of Bohm's Quantum Theory to Include Non-Gradient Potentials and the Production of Nanometer Vortices." http://www22.pair.com/csdc/ pdf/bohmplus.pdf

Klein, Oscar. 1926. "Quantum Theory and Five-Dimensional Theory of Relativity." *Zeitschrift für Physik* 37:895–906.

Kline, Morris. 1980. *Mathematics: The Loss of Certainty*. New York: Oxford University Press.

Kockelmans, Joseph J. 2002. *Ideas for a Hermeneutic Phenomenology of the Natural Sciences*. Dordrecht: Kluwer.

Kuhn, Thomas S. 1962. *The Structure of Scientific Revolutions*. Chicago: University of Chicago Press.

Lacan, Jacques. 1953. "Some Reflections on the Ego." *International Journal of Psychoanalysis* 34:11–17.

Lapedes, Daniel N., ed. 1978. *McGraw-Hill Dictionary of Physics and Mathematics*. New York: McGraw-Hill.

Leder, Drew. 1990. *The Absent Body*. Chicago: University of Chicago Press.

Linde, Andrei. 2005. "Prospects of Inflation." *Physica Scripta* T117:40–48.

Liu, Chuang. 2002. "The Meaning of Spontaneous Symmetry Breaking (I)." http://philsci-archive.pitt.edu/archive/00000563/

Macquarrie, John. 1968. *Martin Heidgger*. Richmond, VA: John Knox.

Mandelbrot, Benoit. 1977. *Fractals*. San Francisco: Freeman.

Matsuno, Koichiro. 1995. "Use of Natural Languages in Modelling Evolutionary Processes." *Proceedings of the 14th International Congress of Cybernetics* (International Association of Cybernetics, Namur, Belgium), 477–82.

———. 1998. "Space-Time Framework of Internal Measurement." In *Computing Anticipatory Systems, AIP Conference Proceedings 437*, edited by D. M. Dubois, 101–15. Woodbury, NY: American Institute of Physics.

Menger, Karl. 1940. "Topology without Points." *Rice Institute Pamphlet* 27:80–107.

Merleau-Ponty, Maurice. 1945/1962. *The Phenomenology of Perception*, translated by Colin Smith. London: Routledge and Kegan Paul.

———. 1956–57/2003. "First Course: The Concept of Nature." In *Nature*, translated by Robert Vallier, 3–122. Evanston, IL: Northwestern University Press.

———. 1964. "Eye and Mind." In *The Primacy of Perception*, edited by James M. Edie, 159–90. Evanston, IL: Northwestern University Press.

———. 1968. *The Visible and the Invisible*, translated by Alphonso Lingis. Evanston, IL: Northwestern University Press.

Musès, Charles. 1968. "Hypernumber and Metadimension Theory." *Journal of Consciousness Studies* 1:29–48.

———. 1975. "Fractional Dimensions and Their Experiential Meaning." *Mathematical/Physical Correspondence* 11:17–21.

———.1976. "Applied Hypernumbers: Computational Concepts." *Applied Mathematics and Computation* 3:211–26.

———. 1977. "Explorations in Mathematics." *Impact of Science on Society* 27:67–85.

Ouspensky, P. D. 1970. *Tertium Organum*. New York: Vintage.

Pagels, Heinz R. 1985. *Perfect Symmetry*. New York: Bantam.

Penrose, Roger. 1960. "A Spinor Approach to General Relativity." *Annals of Physics* 10: 171–201.

———. 1967. "Twistor Algebra." *Journal of Mathematical Physics* 8:345–66.

————. 1971. "Angular Momentum: An Approach to Combinatorial Space-Time." In *Quantum Theory and Beyond*, edited by Ted Bastin, 151–80. Cambridge: Cambridge University Press.

————. 1987. "On the Origins of Twistor Theory." In *Gravitation and Geometry: A Volume in Honour of I. Robinson*, edited by W. Rindler and A. Trautman, 341–61. Naples: Bibliopolis. (http://users.ox.ac.uk/~tweb/00001/index.shtml)

Persichetti, Vincent. 1961. *Twentieth-Century Harmony.* New York: W.W. Norton.

Plato. 1965. *Timaeus and Critias*, translated by Desmond Lee. New York: Penguin.

Pylkkö, Pauli. 1998. *The Aconceptual Mind.* Amsterdam and Philadelphia: John Benjamins.

Rosen, Steven M. 1975. "Synsymmetry." *Scientia* 110:539–49.

————. 1983. "The Concept of the Infinite and the Crisis in Modern Physics." *Speculations in Science and Technology* 6:413–25.

————. 1988. "A Neo-Intuitive Proposal for Kaluza-Klein Unification." *Foundations of Physics* 18:1093–1139.

————. 1994. *Science, Paradox, and the Moebius Principle.* Albany: State University of New York Press.

————. 1995. "Pouring Old Wine Into a New Bottle." In *The Interactive Field in Analysis*, edited by Murray Stein, 121–41. Wilmette, IL: Chiron.

————. 1996. "How Intimate the Flesh?" Presented at Twenty-First Annual International Conference of the Merleau-Ponty Circle, University of Memphis, Memphis, TN, September 20.

————. 1997. "Wholeness as the Body of Paradox." *Journal of Mind and Behavior* 18: 391–423.

————. 2004. *Dimensions of Apeiron.* Amsterdam: Editions Rodopi.

————. 2006. *Topologies of the Flesh.* Athens, OH: Ohio University Press.

Rouse, Joseph. 2003. *How Scientific Practices Matter.* Chicago: University of Chicago Press.

Rucker, Rudolph. 1977. *Geometry, Relativity, and the Fourth Dimension.* New York: Dover.

Ryan, Paul. 1993. *Video Mind/Earth Mind: Art, Communications, and Ecology.* New York: Peter Lang.

Sabelli, Hector. 2005. *Bios: A Study of Creation.* Hackensack, NJ: World Scientific.

Sachs, Mendel. 1999. "Fundamental Conflicts in Modern Physics and Cosmology." *Frontier Perspectives* 8:13–19.

Sartre, Jean-Paul. 1943/1956. *Being and Nothingness: An Essay on Phenomenological Ontology*, translated by Hazel E. Barnes. New York: Philosophical Library.

Scharff, Robert C. and Val Dusek, eds. 2003. *Philosophy of Technology.* Oxford: Blackwell.

Scherk, Joël, and John H. Schwartz. 1975. "Dual Field Theory of Quarks and Gluons." *Physics Letters* 57B:463–66.

Schroedinger, Erwin. 1961. "Our Image of Matter." In *On Modern Physics*, 45–66. New York: Clarkson N. Potter.

Schumacher, John. 1989. *Human Posture*. Albany: State University of New York Press.

Schwartz-Salant, Nathan. 2007. *The Black Nightgown*. Wilmette, IL: Chiron.

Schwenk, Theodor. 1965. *Sensitive Chaos*. London: Rudolf Steiner.

Sheets-Johnstone, Maxine. 1990. *The Roots of Thinking*. Philadelphia: Temple University Press.

Smolin, Lee. 1997. *The Life of the Cosmos*. Oxford: Oxford University Press.

Snow, C. P. 1959. *Two Cultures and the Scientific Revolution*. Cambridge: Cambridge University Press.

Spiegelberg, Herbert. 1982. *The Phenomenological Movement*. The Hague: Martinus Nijhoff.

Stapp, Henry. 1979. "Whiteheadian Approach to Quantum Theory and the Generalized Bell's Theorem." *Foundations of Physics* 9:1–25.

Steinhardt, Paul J., and Neil Turok. 2002. "A Cyclic Model of the Universe." *Science* 296:1436–39.

Suvin, Darko. 1997. "On Cognitive Emotions and Topological Imagination." Background paper for After-Postmodernism Conference, University of Chicago, November 14–16. http://www.focusing.org/apm_papers/suvin.html.

Tanabe, Hajime. 1986. *Philosophy as Metanoetics*, translated by Takeuchi Yoshinori. Berkeley: University of California Press.

Uchii, Soshichi. 1998. "Philosophy of Science in Japan." http://www.bun.kyoto-u.ac.jp/~suchii/philsci_j7.html

Vaughan, Frances. 1979. *Awakening Intuition*. New York: Doubleday/Anchor.

Whitehead, Alfred North. 1978. *Process and Reality*. New York: Free Press.

Willard, Dallas. (n.d.) "Historical and Philosophical Foundations of Phenomenology." http://www.dwillard.org/articles/artview.asp?artID=32

Witten, Edward. 1981. "The Search for a Realistic Kaluza-Klein Theory." *Nuclear Physics* B186:412–28.

———. 1995. "String Theory Dynamics in Various Dimensions." *Nuclear Physics* B443: 85–126.

———. 2004. "Perturbative Gauge Theory as a String Theory in Twistor Space." Cornell University online archive (arXiv): hep-th/0312171, v2, October 6.

Young, Arthur. 1976. *The Reflexive Universe*. New York: Delacorte.

Index

SERIES ON KNOTS AND EVERYTHING

Editor-in-charge: Louis H. Kauffman *(Univ. of Illinois, Chicago)*

The Series on Knots and Everything: is a book series polarized around the theory of knots. Volume 1 in the series is Louis H Kauffman's Knots and Physics.

One purpose of this series is to continue the exploration of many of the themes indicated in Volume 1. These themes reach out beyond knot theory into physics, mathematics, logic, linguistics, philosophy, biology and practical experience. All of these outreaches have relations with knot theory when knot theory is regarded as a pivot or meeting place for apparently separate ideas. Knots act as such a pivotal place. We do not fully understand why this is so. The series represents stages in the exploration of this nexus.

Details of the titles in this series to date give a picture of the enterprise.

Published: